# Nonlocal Variations *and* Local Invariance *of* Fields

# Modern Analytic *and* Computational Methods *in* Science *and* Mathematics

A GROUP OF MONOGRAPHS AND ADVANCED TEXTBOOKS

*Richard Bellman,* EDITOR
University of Southern California

## *Published*

1. R. E. Bellman, R. E. Kalaba, and Marcia C. Prestrud, Invariant Imbedding and Radiative Transfer in Slabs of Finite Thickness, 1963

2. R. E. Bellman, Harriet H. Kagiwada, and Marcia C. Prestrud, Invariant Imbedding and Time-Dependent Transport Processes, 1964

3. R. E. Bellman and R. E. Kalaba, Quasilinearization and Nonlinear Boundary-Value Problems, 1965

4. R. E. Bellman, R. E. Kalaba, and Jo Ann Lockett, Numerical Inversion of the Laplace Transform: Applications to Biology, Economics, Engineering, and Physics, 1966

5. S. G. Mikhlin and K. L. Smolitskiy, Approximate Methods for Solution of Differential and Integral Equations, 1967

6. R. N. Adams and E. D. Denman, Wave Propagation and Turbulent Media, 1966

7. R. L. Stratonovich, Conditional Markov Processes and Their Application to the Theory of Optimal Control, 1968

8. A. G. Ivakhnenko and V. G. Lapa, Cybernetics and Forecasting Techniques, 1967

9. G. A. Chebotarev, Analytical and Numerical Methods of Celestial Mechanics, 1967

10. S. F. Feshchenko, N. I. Shkil', and L. D. Nikolenko, Asymptotic Methods in the Theory of Linear Differential Equations, 1967

12. R. E. Larson, State Increment Dynamic Programming, 1968

13. J. Kowalik and M. R. Osborne, Methods for Unconstrained Optimization Problems, 1968

14. S. J. Yakowitz, Mathematics of Adaptive Control Processes, 1969

16. D. U. von Rosenberg, Methods for the Numerical Solution of Partial Differential Equations, 1969

18. R. Lattès and J.-L. Lions, The Method of Quasi-Reversibility: Applications to Partial Differential Equations. Translated from the French edition and edited by Richard Bellman, 1969

19. D. G. B. Edelen, Nonlocal Variations and Local Invariance of Fields, 1969

## *In Preparation*

11. A. G. Butkovskiy, Distributed Control Systems

15. S. K. Srinivasan, Stochastic Theory and Cascade Processes

17. R. B. Banerji, Theory of Problem Solving: An Approach to Artificial Intelligence

20. J. R. Radbill and G. A. McCue, Quasilinearization and Nonlinear Problems in Fluid and Orbital Mechanics

21. W. Squire, Integration for Engineers and Scientists

22. T. Parthasarathy and T. E. S. Raghavan, Some Topics in Two-Person Games

23. T. Hacker, Flight Stability and Control

# Nonlocal Variations *and* Local Invariance *of* Fields

Dominic G. B. Edelen
*Purdue University,* Lafayette, Indiana

American Elsevier
Publishing Company, Inc.
NEW YORK · 1969

AMERICAN ELSEVIER PUBLISHING COMPANY, INC.
52 Vanderbilt Avenue, New York, N.Y. 10017

ELSEVIER PUBLISHING COMPANY, LTD.
Barking, Essex, England

ELSEVIER PUBLISHING COMPANY
335 Jan Van Galenstraat, P. O. Box 211
Amsterdam, The Netherlands

Standard Book Number 444-00054-2

Library of Congress Card Number 70-75525



*Manufactured in the United States of America*

*To the*

GENTLE

PEOPLE

*of*

Tidewater,

Maryland

# Contents

## CHAPTER 3

## Geometric Objects and Lie Derivatives



## CHAPTER 4

## Invariance Considerations

# Preface

THE trend toward classification and study of nonlocalizable properties of physical processes is unmistakable. Indeed, it is a logical consequence of the ever increasing demand for precision in the quantification and description of physical systems. New demands are thus placed on the practitioners of the sciences, and new techniques are required since the very essence of nonlocalizability denies the ability to describe systems solely in terms of differential equations. The intent of this book is to equip the reader with some of the techniques and methods that are required in these new areas and to render this information accessible to the second-year graduate student. Accordingly, an attempt has been made to make the text as self-contained as possible.

The basic techniques of the text are obtained from appropriate extensions of the classical variational methods since these methods provide an underlying discipline for the description and analysis of localizable phenomena. These extensions provide for the consideration of functionals of functionals and lead to Euler equations that are integro-differential in nature. The second half of the book examines the questions of invariance of variational statements under a wide range of transformations and functional substitutions. After an analysis of the general case the investigation is restricted to localizable structures. This restriction was found advisable in view of the numerous unresolved questions that occur even in the localizable case and in view of the fact that the nonlocalizable case does not provide a basis for direct inference. The inclusion of the material on invariance is deemed necessary since it provides the basic structure whereby systems of significant complexity can be modeled from elementary information concerning their generalized symmetries. The theory of geometric object fields and Lie derivatives is introduced since it provides significant simplification of many of the calculations and provides a structure of sufficient generality to encompass almost any situation.

Only questions of stationarity of functionals are considered in this book. The reason for this restriction is that the equations governing most physical systems cannot be obtained from variational statements involving functionals that attain maximal or minimal values—most functionals of physical interest (generalized action functionals) exhibit conjugate points or other difficulties such that their stationary points correspond to points of inflection or saddle points in the ordinary calculus.

The setting for the discussion is a space of independent variables of $n$ dimensions and a collection of functions defined over a given compact set of this space. Chapter 1 confines the discussion to the case where each collection of functions has only one element, and the collections form a function space of differentiable functions. The calculus of variations is then developed in terms of lagrangian functions which contain functionals as arguments. The resulting Euler equations (the necessary conditions for stationarity of the integral of the lagrangian function) thus involve both integral and differential operators. This leads to a consistent means of examining integral, differential, and integro–differential equations on an equal basis. The most general lagrangian function which leads to linear Euler equations is examined. The results of this examination give a direct answer to the variational embedding problem, namely, when can a given linear equation be embedded in a variational statement involving only one unknown function. The analysis is given in some detail so that the reader may realize just how rich a structure may be obtained from the variational calculus of functionals of functionals. Inclusion of functionals as arguments of the lagrangian function significantly increases the scope of the null class of the Euler-Lagrange operator (the collection of all lagrangian functions such that every function in the function space renders the functional formed from such a lagrangian function stationary in value). Explicit characterization of the null class is given together with a study of the additive augmentation of the original functional by functionals whose support is the boundary of the compact set of definition of the original functional. This leads to the study of natural boundary conditions and to the fact that non-natural boundary conditions can be obtained as natural ones by adding to the original lagrangian function an appropriate element of the null class of the Euler-Lagrange operator.

Chapter 2 extends the results of the first chapter to the case in which there are several dependent functions to be chosen in order to render a given functional stationary in value. This study leads to the fundamental adjoint theorem and to the variational embedding of any system of second-order equations in a variational statement. The form of this embedding is such that a simple argument leads to a direct method of obtaining necessary conditions for the existence of solutions to the given system. This method is the same for differential, integral, and integro–differential equations and leads to the classical results. The point of interest is that the necessary conditions obtained in this way for the existence of a potential for a given vector field and the conditions for existence of solutions to integral equations of the first kind, to give two examples, are obtained in exactly the same way. The usual dichotomy between these two subjects is thus avoided. Momentum-energy complexes are then developed for functionals of functionals, and the chapter concludes with a demonstration of the invariance of the Euler equations.

Chapter 3 is devoted to the study of geometric object fields and to Lie derivatives of such fields. The purpose of this chapter is to establish the necessary groundwork for Chapter 4.

Invariance properties of functionals are developed in Chapter 4. The variations induced in functionals of functionals by coordinate transformations and function variations are first computed, and these are combined to obtain the induced variation under point transformations. Consideration is then restricted to localizable problems in which a number of different kinds of invariance are considered and their defining requirements are obtained. The problem is then reversed so that we look for all lagrangian functions which admit a given type of invariance for a given group of transformations. This leads to a system of partial differential equations whose general solution gives us the possible lagrangian functions with the desired properties.

The text ends with an Appendix on the problems of stationarizing functionals of functionals with constraints. The results obtained here show again that the calculus of functionals of functionals provides a more natural setting since the various cases of classical Lagrange multiplier problems all reduce to a simple extension of the associated advanced calculus problem. The general case allows for significantly more general situations than the corresponding classical ones.

In most respects the notation is standard. The summation convention is assumed throughout. Thus, if any indexed quantity has the same index letter appearing more than once, the index is to be summed over the range of the index involved. This applies to lower case Latin indices, lower case Greek indices, and upper case Greek indices.

The correct assignment of credits in a work of this nature is almost impossible, and so I have taken the easy way out and not made the attempt. Approximately one-third of the material, mostly that dealing with functionals of functionals, is new as far as I can determine. With the vast amount of literature today, however, I can only claim originality for any part of the material on a conjectural basis.

I would be sincerely remiss if I did not take this opportunity to express my appreciation to the students of the MA 692-C class at Purdue University who bore up most admirably under the multifarious iteration processes that were necessary in wresting this text into its final form. Without their patience and suggestions, the work would never have been finished. I also owe a debt of gratitude to Mrs. Helen Brickler for her patience in the preparation of the final manuscript.

Dominic G. B. Edelen*

Purdue University
Lafayette, Indiana
December 1968

* Present Address:
Center for the Application of Mathematics
Lehigh University
Bethlehem, Pennsylvania

# 1 Variations with One Dependent Function

## 1.1. FUNCTIONALS

Let $D$ be an open, arcwise simply connected point set in $n$-dimensional number space $X_n$, and let $D^*$ denote the closure of $D$ with respect to $X_n$. The set $D^*$ will always be assumed to be *compact*. The boundary of $D$, namely, $D^* - D$, where $-$ denotes the set-theoretic difference, is denoted by $\partial D$. Let $(x)$ be a coordinate system on $D^*$. The "coordinate differential volume element" of $D^*$ with respect to the $(x)$-coordinate system is denoted by $dV(x)$, and the "coordinate differential element of directed surface" of $\partial D$ with respect to the $(x)$-coordinate system is denoted by $dS_i(x)$.

Let $K$ be a class of functions whose domains of definition contain the set $D^*$. A *functional* is a correspondence that assigns a definite number to each element of the class $K$ when the domain of each element (each function) is restricted to the common domain $D^*$; a mapping of $K$ into the real numbers. A functional may thus be thought of as a kind of function where the independent variable is iteself a function (an element of $K$).

The kinds of functionals of primary interest to us are obtained as follows. Let $F(x^k, y)$ be a continuous function of its $n+1$ arguments $x^1, x^2, \ldots, x^n, y$, The quantity $J[y]$, defined by

$$J[y] = \int_{D^*} F\left[x^k, y(x^k)\right] dV(x),$$

where $y(x^k)$ is any continuous function defined on $D^*$, is a functional on the class of continuous functions defined on $D^*$. More generally, if $F(x^k, y, z^k)$ is a continous function of its $2n+1$ arguments, then the quantity

$$J[y] = \int_{D^*} F\left[x^k, y(x^k), \partial_m y(x^k)\right] dV(x) \; , \qquad \partial_m y(x^k) = \frac{\partial y(x^k)}{\partial x^m}$$

is a functional on the class of $C^1$ functions defined over $D^*$.

The value of $J[y]$ in each of the above examples will change when different functions are selected from the class $K$. What we are interested in is just how such changes occur and, in particular, if an element $y_0$ of $K$ can be found such that the values of $J[y]$ and $J[y_0]$ will be identical whenever the element $y$ is near the element $y_0$ in $K$. Such problems require a precise definition of what is meant by two elements of $K$ being near each other. In the ordinary calculus, the distances between points in a plane is well defined, and what is meant in that context by two points being near each other is known. In dealing with functionals, however, we do not have the advantage of *a prior* definition. Since a functional is like a function of a function, we will want to examine continuity and differentiability of functionals in a fashion analogous to the calculus of ordinary functions. These concepts also require the idea of closeness in the class $K$.

## 1.2. FUNCTION SPACES

The concept of continuity plays as important a role in the study of functionals as it does in the study of functions. Now, continuity involves the notion of closeness; and therefore, this concept must be introduced into the context of a class $K$ of functions if we are to handle functionals successfully. The introduction of closeness, in turn, requires a certain structure on the elements of $K$.

By a *linear space*, we mean a set $\mathfrak{R}$ (or possibly a class) of elements $x, y, z, \ldots$ of any kind; an operation $+$ on $\mathfrak{R} \times \mathfrak{R}$ to $\mathfrak{R}$, and an operation $\cdot$ on $E_1 \times \mathfrak{R}$ to $\mathfrak{R}$ ($E_1$ = one-dimensional number space) such that

1. $x + y = y + x$,
2. $(x + y) + z = x + (y + z)$,
3. there is an element $\emptyset$ (the identity element of addition) such that $x = \emptyset + x$ for every $x \in \mathfrak{R}$,
4. for each $x \in \mathfrak{R}$ there exists an element $-x$ such that $x + (-x) = \emptyset$,
5. $1 \cdot x = x$,
6. $a \cdot (b \cdot x) = ab \cdot x$,
7. $(a + b) \cdot x = a \cdot x + b \cdot x$,
8. $a \cdot (x + y) = a \cdot x + a \cdot y$.

A linear space $\mathfrak{R}$ is said to be a *normed* linear space, if each element $x \in \mathfrak{R}$ is assigned a nonnegative number $\|x\|$, called the *norm* of $x$, such that

1. $\|x\| = 0$ if and only if $x = \emptyset$,
2. $\|a \cdot x\| = |a| \, \|x\|$,
3. $\|x + y\| \leq \|x\| + \|y\|$.

We can speak of the distance between two elements $x$ and $y$ of a normed linear space by simply defining the distance to be the norm of the sum of $x$ and $-y$. Accordingly, two elements of a normed linear space are near one another if $\|x - y\| < \delta$ for $\delta$ sufficiently small. The reader should note that a norm on a linear space of functions is a functional whose range is the nonnegative numbers.

The following normed linear spaces are important:

1. The space $\mathcal{C}(D^*)$ consisting of all continuous functions $y(x^k)$ defined on $D^*$. By addition of elements of $\mathcal{C}(D^*)$ and multiplication of elements of $\mathcal{C}(D^*)$ by numbers, we mean ordinary addition of functions and multiplication of functions by numbers. The norm is defined by

$$\|y\| = \max_{x \in D^*} |y(x^k)|. \tag{1.1}$$

2. The space $\mathfrak{D}_1(D^*)$ consisting of all continuous functions $y(x^k)$ defined on $D^*$ with continuous derivatives of the first order with respect to each of the $n$ coordinates of $D^*$. Addition and multiplication by numbers are the same as for $\mathcal{C}(D^*)$. The norm is defined by

$$\|y\| = \max_{x \in D^*} |y(x^k)| + \sum_{i=1}^{n} \max_{x \in D^*} |\partial_i y(x^k)|. \tag{1.2}$$

If a sequence of elements $y_k$ of $\mathcal{C}(D^*)$ converges to an element $y \in \mathcal{C}(D^*)$ in the sense that

$$\lim_{k \to \infty} \|y_k - y\| = 0 \tag{1.3}$$

(convergence in $\mathcal{C}(D^*)$-norm), then the sequence is uniformly convergent for all $x^k$ in $D^*$. The norm (1.1) on $\mathcal{C}(D^*)$ is thus referred to as the norm of uniform convergence. If a sequence of elements

$y_k$ of $\mathfrak{D}_1(D^*)$ converges to an element $y$ of $\mathfrak{D}_1(D^*)$ in the norm on $\mathfrak{D}_1(D^*)$, then the sequence and its first partial derivatives converge unformly to $y$ and its first partial derivatives, respectively, throughout $D^*$.

The concept of continuity follows naturally from the discussion of "nearness." Let $\mathfrak{R}$ be a normed linear space of functions and let $J[y]$ be a functional defined on $\mathfrak{R}$. The functional $J[y]$ is said to be *continuous at the element* $y_0$ of $\mathfrak{R}$ if, for any given $\epsilon > 0$, there exists a positive number $\delta$ such that

$$|J[y] - J[y_0]| < \epsilon \tag{1.4}$$

whenever $\|y - y_0\| < \delta$ If $J[y]$ is continuous at every element of $\mathfrak{R}$, then $J[y]$ is said to be a *continuous functional on* $\mathfrak{R}$. If (1.4) is replaced by

$$J[y] - J[y_0] > -\epsilon, \tag{1.5}$$

then $J[y]$ is said to be *lower semicontinuous at* $y_0$. If (1.4) is replaced by

$$J[y] - J[y_0] < \epsilon, \tag{1.6}$$

then $J[y]$ is said to be *upper semicontinuous at* $y_0$.

Let $G(x^m, z)$ be continuous in its $n+1$ arguments for all $\{x^m\}$ in $D^*$ and all real $z$, and set

$$J[y] = \int_{D^*} G[x^m, y(x^m)]\, dV(x). \tag{1.7}$$

We then have

$$|J[y] - J[y_0]| \leq \int_{D^*} |G[x^m, y(x^m)] - G[x^m, y_0(x^m)]|\, dV(x). \tag{1.8}$$

For the normed linear space of functions, $\mathfrak{R}$, we take $\mathcal{C}(D^*)$. Since $D^*$ is compact, there exists a number $M$ such that $|y_0(x^m)| \leq M$, for all $\{x^m\} \in D^*$. For $y(x^m)$ such that $\|y - y_0\| < \delta$, we have $|y(x^m) - y_0(x^m)| < \delta$ for all $\{x^m\} \in D^*$, since $\mathcal{C}(D^*)$ has the uniform convergence norm. Thus, $|y(x^m)| < M + \delta$ for all $\{x^m\} \in D^*$. If $I$ denotes the closed segment of the real line $[-M - \delta, M + \delta]$ then $D^*XI$ is compact, and hence the continuity of $G(x^m, z)$ on $D^* X$ (reals) implies the uniform continuity of $G(x^m, z)$ on $D^* X I$. It thus follows that, for any given $\epsilon' > 0$, $|G(x^m, z) - G(x^m, z_0)| < \epsilon'$ whenever $|z - z_0| < \delta$ for all $\{x^m\} \in D^*$ and all $z_0$ in $[-M, M]$. Identifying $z_0$ with $y_0(x^m)$ and $z$ with $y(x^m)$ for any fixed $\{x^m\} \in D^*$, we obtain

$$|G[x^m, y(x^m)] - G[x^m, y_0(x^m)]| < \epsilon'$$

for each $\{x^m\} \in D^*$ whenever $\| y - y_0 \| < \delta$ and $\| y_0 \| < M$ and $\epsilon'$ and $\delta$ are independent of the choice of $\{x^m\} \in D^*$. We accordingly obtain

$$|J[y] - J[y_0]| < \int_{D^*} \epsilon' dV(x) = \epsilon' V(D^*),$$

where $V(D^*)$ denotes the volume of $D^*$ and is finite since $D^*$ is compact. Thus, if we set $\epsilon = \epsilon' V(D^*)$, then $|J[y] - J[y_0]| < \epsilon$ whenever $\| y - y_0 \| < \delta$ and $\| y_0 \| \leq M$. Therefore $J[y]$ is continuous at $y_0$. It is now clear that $J[y]$ is continuous on $\mathcal{C}(D^*)$ since any element of $\mathcal{C}(D^*)$ is bounded, i.e., *continuity of* $G(x^m, z)$ *implies continuity of* $J[y]$ *when this functional is defined by* (1.7). Problem 1.2 gives the extension to $\mathfrak{D}_1(D^*)$.

There are obviously weaker conditions on $G(x^m, z)$ and different normed linear spaces of functions that lead to continuity of functionals. The literature abounds with examples. Suppose that $J[y]$ is defined by (1.7) and that

$$|G[x^m, y(x^m)] - G[x^m, y_0(x^m)]| \leq M(x^m)\, f\Big[|y(x^m) - y_0(x^m)|\Big],$$

where $M(x^m) > 0$ and Riemann integrable on $D^*$ with $\int_{D^*} M(x^m)\, dV(x) = M$ and $f(z)$ is bounded, monotone increasing, and continuous in $z$ with $f(0) = 0$. In this case, if $y(x^m)$ and $y_0(x^m)$ are elements of $\mathcal{C}(D^*)$, then $\| y - y_0 \| < \delta$ implies $|y(x^m) - y_0(x^m)| < \delta$ for all $\{x^m\} \in D^*$, therefore,

$$|J[y] - J[y_0]| < f(\delta) M.$$

It then follows that $J[y]$ is continuous at $y_0$ on setting $\delta = f^{-1}(\epsilon/M)$.

## 1.3. FUNDAMENTAL LEMMAS

Let $z(x^m)$ and $z^k(x^m)$, $k = 1, \ldots, n$ be a fixed collection of $n + 1$ elements of $\mathfrak{D}_1(D^*)$. Then the integral

$$J[y] = \int_{D^*} (yz + z^k \partial_k y)\, dV(x) \tag{1.9}$$

defines a functional on $\mathfrak{D}_1(D^*)$ which is continuous and linear. Suppose that $J[y]$ vanishes for all $y(x^m)$ belonging to some class of functions in $\mathfrak{D}_1(D^*)$. If this is to be the case, the functions $z$ and $z^k$ must possess certain specific properties.

**Lemma 1.1.** *If $z(x^m)$ is an element of $\mathcal{C}(D^*)$ and if*

$$J[y] = \int_{D^*} y(x^m)\, z(x^m)\, dV(x)$$

*vanishes for every $y(x^m) \in \mathcal{C}(D^*)$ such that $y(x^m)\Big|_{\partial D} = 0$, then $z(x^m) = 0$ on $D^*$.*

**Proof.** Suppose that $z(x^m)$ is nonzero at some point $x_0$ interior to $D^*$, and, for definiteness, assume that $z(x_0^m) > 0$. Since $z(x^m) \in \mathcal{C}(D^*)$, there exists a sphere $s(r)$ interior to $D^*$ with center $x_0$ and radius $r$ such that $z(x^m)$ is positive for all points interior to and on $s(r)$. Denote by $\bar{s}(r)$ the set of all points of $D^*$ that lie in the interior of $s(r)$ or on $s(r)$. Take $y(x^m)$ to be given by

$$y(x^m) = \begin{cases} r^2 - \sum_{k=1}^{n} \left(x^k - x_0^k\right)^2 & \text{for all } x \in \bar{s}(r) \\ 0 & \text{for all } x \notin \bar{s}(r). \end{cases}$$

Then $y(x^m) \in \mathcal{C}(D^*)$ and vanishes on the boundary of $D^*$. Since this $y(x^m)$ satisfies the conditions placed on $y(x^m)$ by the hypothesis of Lemma 1.1 and since simple calculation gives

$$J[y] = \int_{\bar{s}(r)} y(x^m)\, z(x^m)\, dV(x) > 0,$$

we have a contradiction. Hence the assumption is false that there is a point $x_0$ interior to $D^*$ for which $z\left(x_0^m\right) \neq 0$. Thus, $z(x^m)$ vanishes at all interior points of $D^*$, and hence $z(x^m)$ vanishes throughout $D^*$ by the assumed continuity of $z(x^m)$.

The hypothesis of Lemma 1.1 required $J[y]$ to vanish for all elements of $\mathcal{C}(D^*)$ that vanish on the boundary of $D^*$. Since every element of $\mathfrak{D}_1(D^*)$ is an element of $\mathcal{C}(D^*)$, while the converse is not true, we can obtain a stronger result by weakening the hypothesis so that $y$ belongs to $\mathfrak{D}_1(D^*)$ rather than to $\mathcal{C}(D^*)$.

**Lemma 1.2.** If $z(x^k)$ *is an element of* $\mathcal{C}(D^*)$ *and if*

$$J[y] = \int_{D^*} y(x^m)\, z(x^m)\, dV(x)$$

*vanishes for every* $y(x^m) \in \mathfrak{D}_1(D^*)$ *which vanishes on the boundary of* $D^*$, *then* $z(x^m) = 0$ *on* $D^*$.

**Proof.** The proof is the same as for Lemma 1.1 with

$$y(x^m) = \begin{cases} \left[ r^2 - \sum_{k=1}^{n} \left(x^k - x_0^k\right)^2 \right]^2 & \text{for } x \in \bar{s}(r) \\ 0 & \text{for } x \notin \bar{s}(r) \end{cases}$$

since $y(x^m)$ is an element of $\mathfrak{D}_1(D^*)$ and vanishes on the boundary of $D^*$.

**Lemma 1.3.** *If* $z(x^m) \in \mathcal{C}(D^*)$ *and if*

$$J[y] = \int_{D^*} y(x^m)\, z(x^m)\, dV(x)$$

*vanishes for all functions* $y(x^m) \in \mathcal{C}(D^*)$ *or all* $y(x^m) \in \mathfrak{D}_1(D^*)$, *then* $z(x^m) = 0$ *on* $D^*$.

**Proof.** The proof is the same as for Lemma 1.1 or 1.2 with the exception that we no longer require $y(x^m)$ to vanish on the boundary of $D^*$. This is easily done by defining $y(x^m)$ as in Lemma 1.1 or 1.2 and noting that $\bar{s}(r)$ is no longer required to lie completely within $D^*$. In this case we have

$$J[y] = \int_{\bar{s} \cap D^*} y(x^m)\, z(x^m)\, dV(x) > 0\,.$$

**Lemma 1.4.** *If* $z(x^m) \in \mathcal{C}(D^*)$ *and* $z^k(x^m) \in \mathfrak{D}_1(D^*)$ *for each* $k = 1, \ldots, n$, $\partial_k z^k \in \mathcal{C}(D^*)$ *and if*

$$J[y] = \int_{D^*} \left(zy + z^k \partial_k y\right) dV(x)$$

*vanishes for all* $y(x^m) \in \mathfrak{D}_1(D^*)$ *such that* $y(x^m)$ *vanishes on* $\partial D^*$*, then*

$$z = \partial_k z^k$$

*holds on* $D^*$.

**Proof.** When the divergence theorem is used, we have

$$J[y] = \int_{D^*} \left(z - \partial_k z^k\right) y \, dV(x) + \int_{\partial D^*} y \, z^k \, dS_k(x) .$$

Since $y(x^m)$ vanishes on the boundary of $D^*$, the surface integral vanishes. The result then follows on applying Lemma 1.2.

## 1.4. FUNCTIONAL DERIVATIVES, VARIATIONS

One way of viewing derivatives in the ordinary calculus is to consider the approximating linear function of a given function. Since the linearity of the linear approximation is handled in a straightforward fashion in a linear space, it is this approach that is used in the context of functionals. This necessitates a precise definition of a linear functional. A functional $J[y]$ on a normed linear space of function $\mathfrak{R}$ is said to be a *linear functional* if an only if $J[y]$ is continuous on $\mathfrak{R}$ and such that

$$J[a \cdot y_1 + b \cdot y_2] = aJ[y_1] + bJ[y_2]$$

holds for all real numbers $a$ and $b$ and for all $y_1, y_2 \in \mathfrak{R}$ If we set $a = b = 0$ then $a \cdot y_1 + b \cdot y_2 = \emptyset$; and we obtain $J[\emptyset] = 0$. A linear functional is thus both linear and homogeneous.

Let $J[y]$ be a functional defined on a normed linear space $\mathfrak{R}$, and let $\Delta J[y;h]$ be defined by

$$\Delta J[y;h] = J[y+h] - J[y], \qquad (1.10)$$

where $h \in \mathfrak{R}$. If $y_0$ is a fixed element of $\mathfrak{R}$ and if there exists a functional $j[y_0;h]$ which is a *linear functional* of $h$ such that

$$\Delta J[y_0;h] = j[y_0;h] + o(\|h\|), \tag{1.11}$$

where $o(\|h\|)$ denotes any functional $g[y_0;h]$ such that

$$\lim_{\|h\|\to 0}\left(\frac{g[y_0;h]}{\|h\|}\right) = 0,$$

then $J[y]$ is said to be *differentiable* at $y = y_0$ and $j[y_0;h]$ is said to be the *differential* of $J[y]$ at $y = y_0$. Suppose that $h = \lambda\cdot g$, where $g$ is any element of $\mathfrak{R}$ such that $\|g\| = G$ for given $G$, then $\|h\| = \lambda G$. Thus, it follows that the norm of $h$ tends to zero as $\lambda$ tends to zero. We then have

$$\Delta J[y_0;h] = J[y_0 + \lambda\cdot g] - J[y_0].$$

If $J[y]$ is differentiable at $y_0$, then

$$\begin{aligned}\Delta J[y_0;h] &= j[y_0;h] + o(\|h\|) = j[y_0;\lambda\cdot g] + o(\lambda G)\\ &= \lambda j[y_0;g] + o(\lambda);\end{aligned}$$

and $g$ is an arbitrary element of $\mathfrak{R}$ to within the constraint $\|g\| = G$.

The value of $j[y_0;h]$ will depend on the choice of the element $h$, just as the value of $f(x+h) - f(x)$ depends on the value of $h$ in the ordinary calculus. The problem here is whether $j[y_0;h]$ is the same linear functional of $h$ as $j[y_0;w]$ is of $w$. In order to establish this as the case, the following lemma is given.

**Lemma 1.5.** *If $j[h]$ is a linear functional on a normed linear space and if*

$$\lim_{\|h\|\to 0}\left(\frac{j[h]}{\|h\|}\right) = 0,$$

*then $j[h] \equiv 0$ on $\mathfrak{R}$.*

**Proof.** Suppose that $j[h_0] \neq 0$ for some $h_0 \in \mathfrak{R}$. Since $j[h]$ is a linear functional, we must have $h_0 \neq \emptyset$; and hence $\|h_0\| \neq 0$ Set

$$h_n = \frac{h_0}{n}, \qquad P = \frac{j[h_0]}{\|h_0\|} \neq 0,$$

then the limit of the norm of $h_n$ tends to zero as $n$ tends to infinity. However, $j[h_n] = j[(1/n)\cdot h_0] = j[h_0]/n$ and $\|h_n\| = \|h_0\|/n$. We thus obtain

$$\lim_{\|h\|\to 0}\left(\frac{j[h]}{\|h\|}\right) = \lim_{n\to\infty}\left(\frac{j[h_n]}{\|h_n\|}\right) = P \neq 0,$$

and the result is established.

**Theorem 1.1.** *The differential of a differentiable functional is a unique functional.*

**Proof.** Suppose that the contrary is true, that is,

$$\Delta J[y_0;h] = j_1[y_0;h] + o(\|h\|),$$

$$\Delta J[y_0;h] = j_2[y_0;h] + o(\|h\|).$$

These relations imply that

$$j_1[y_0;h] - j_2[y_0;h] = o(\|h\|).$$

Now, $j_1$ and $j_2$ are linear functionals of $h$ and hence $j_1 - j_2$ is a linear functional $h$. An application of the Lemma 1.5 gives $j_1 - j_2 = 0$ for all $h \in \mathfrak{R}$, and hence $j_1[y_0;h] = j_2[y_0;h]$ for $h \in \mathfrak{R}$. It thus follows that if $J[y]$ is differentiable at $y_0$, then the linear functional $j[y_0;h]$ of $h$, for which

$$\Delta J[y_0;h] = j[y_0;h] + o(\|h\|)$$

is unique.

In the literature the linear functional $j[y_0;h]$ is usually denoted by $\delta J$ and is referred to as the variation of $J$:

$$\Delta J[y_0;h] = \delta J + o(\|h\|). \tag{1.12}$$

It is also customary to write

$$h = \delta y, \quad y_0 + h = y_0 + \delta y. \tag{1.13}$$

For $h$ and $y_0$ belonging to $\mathfrak{D}_1(D^*)$, we then have

$$\partial_k(y_0 + h) = \partial_k(y_0 + \delta y) = \partial_k y_0 + \partial_k \delta y \equiv \partial_k y_0 + \delta\partial_k y. \tag{1.14}$$

The last equality in (1.14) is a carryover from the literature in which a variation is induced by embedding $y_0(x^m)$ in a one-parameter family of functions $y(x^m, \lambda)$ such that $y(x^m, 0) = y_0(x^m)$, $y(x^m, \lambda) \in \mathfrak{D}_1(D^*)$ for all $\lambda$ in some neighborhood of $\lambda = 0$ and such that the partial derivative of $y(x^m, \lambda)$, with respect to $\lambda$ exists in some neighborhood of $\lambda = 0$ and belongs to $\mathfrak{D}_1(D^*)$. Under these conditions, the following is written,

$$y(x^m, \lambda) = y_0(x^m) + \lambda\hat{\delta}y(x^m) + o(\lambda),$$

$$\hat{\delta}y(x^m) = \left.\frac{\partial y(x^m, \lambda)}{\partial \lambda}\right|_{\lambda=0}$$

The reader should carefully note that $\hat{\delta}y$ obtained in this fashion is related to $\delta y$ by $\delta y = \lambda\hat{\delta}y$. The following observation should be sufficient to achieve a satisfactory compromise between the two methods:

$$j[y_0; \delta y] = j[y_0; h] = j[y_0; \lambda \cdot g] = j[y_0; \lambda \cdot \hat{\delta}y] = \lambda j[y_0; \hat{\delta}y].$$

One additional definition is required. A functional $J[y]$ is said to be *differentiable on a region* $\bar{\mathfrak{R}}$ of a normed linear space of functions $\mathfrak{R}$ if and only if $J[y]$ is differentiable at every element of $\bar{\mathfrak{R}}$.

## 1.5. STATIONARITY AND THE EULER-LAGRANGE OPERATOR

Let $g_a\left[x^k, z^k, y(z^k), \partial_m y(z^k)\right]$, $a = 1, \ldots, q$, be $q$ functions of class $C^2$ in their $3n + 1$ arguments. With these functions, we define new functions of $x^k$ by the relations

$$k_a(x^k; y) = \int_{D^*} g_a\left[x^k, z^k, y(z^k), \partial_m y(z^k)\right] dV(z) \tag{1.15}$$

If we consider $x^k$ as parameters, then (1.15) are simply differentiable functionals on $\mathfrak{D}_1(D^*)$ (see Prob. 1.3). Since we are not primarily concerned with pathology, it is convenient to require that $g_a$ be $C^2$ functions. Once we are finished, the reader can relax this condition if a problem should require it.

Let $L\left[x^k, y(x^k), \partial_m y(x^k), k_a(x^k; y)\right]$ be a function of class $C^2$ in its $2n + q + 1$ arguments. This function will be referred to as the *lagrangian function*, and will serve to define a continuous and differentiable functional on $\mathfrak{D}_1(D^*)$ by the relation

$$J[y] = \int_{D^*} L\left[x^k, y(x^k), \partial_m(x^k), k_a(x^k; y)\right] dV(x). \tag{1.16}$$

Situations will arise in which results obtained by using different lagrangian functions must be compared. The values of $J[y]$ will change when the lagrangian function is changed, and thus, it will be useful to introduce the notation

$$J[y](L) = \int_{D^*} L\left[x^k, y(x^k), \partial_m y(x^k), k_a(x^k; y)\right] dV(x). \tag{1.17}$$

Also the value of $J[y](L)$ will change if $D^*$ is replaced by $d^*$, where $d$ is any open, simply connected subset of $D^*$ of $n$ dimensions. For this purpose, we use the notation

$$J[y](L; d) = \int_{d^*} L\left[x^k, y(x^k), \partial_m y(x^k), k_a(x^k; y)\right] dV(x). \tag{1.18}$$

The continuity and differentiability of $J[y](L; d)$ on $\mathfrak{D}_1(D^*)$ follows immediately from the continuity of $J[y](L; D)$ on $\mathfrak{D}_1(D^*)$. The following convention is of obvious use: $J[y](L) = J[y](L; D)$.

*A point* $y_0 \in \mathfrak{D}_1(D^*)$ *is said to be a stationary point of* $J[y](L)$ *if and only if* $J[y](L)$ *is differentiable at* $y_0$ *and* $\delta J[y_0](L) = 0$ *for all variations* $\delta y$ *which vanish on the boundary of* $D^*$. The set $S(L)$ of $\mathfrak{D}_1(D^*)$, comprised solely of stationary points of $J[y](L)$, is said to be the *stationarization set base* $L$. It is noted that a stationarization set will change whenever the functional form of $L$ is changed; and hence, we must always state what lagrangian function is used in determining a stationarization set. For this reason we have used the notation $S(L)$ since the dependence of $S(L)$ on $L$ must be kept clear at all times. One of the problems we shall consider in Sec. 1.6 is that of whether lagrangian functions can be found for which $S(L)$ is identical with $\mathfrak{D}_1(D^*)$.

Application of the above definition demands the calculation of the linear approximating functional $\delta J[y](L)$. From the definition of $\Delta J$ given in Sec. 1.4 and from (1.17), we have

$$\Delta J[y](L) = \int_{D^*} L\left[x^k, y + \delta y, \partial_m y + \partial_m \delta y, k_a(x^k; y + \delta y)\right] dV(x) - \int_{D^*} L\left[x^k, y, \partial_m y, k_a(x^k; y)\right] dV(x), \tag{1.19}$$

in which $\delta y$ is used in place of the function $h$ used previously. When (1.15) is used, togather with the assumed continuity properties of $g_a$, the following is easily obtained:

$$k_a(x^k; y + \delta y) = k_a(x^k; y) + o(\| \delta y \|)$$
$$+ \int_{D^*} \left\{ \frac{\partial g_a}{\partial y(z)} \delta y(z) + \frac{\partial g_a}{\partial [\partial_m y(z)]} \partial_m \delta y(z) \right\} dV(z) .$$

The integral on the right-hand side of the above equality is the linear approximating functional of the differentiable functional $k_a(x^k; y)$, which involves not only $\delta y(x^k)$, but also $\partial_m \delta y(x^k)$. Ultimately, if we are to obtain explicit conditions for a stationary point, we shall want to be able to apply Lemma 1.2 which requires that only $\delta y(x^k)$ appear, not $\partial_m \delta y(x^k)$. Suppose that we can write

$$\frac{\partial g_a}{\partial [\partial_m y(z)]} \partial_m \delta y(z) = \frac{\partial}{\partial z^m} \left\{ \frac{\partial g_a}{\partial [\partial_m y(z)]} \delta y(z) \right\} - \delta y(z) \frac{\partial}{\partial z^m} \left\{ \frac{\partial g_a}{\partial [\partial_m y(z)]} \right\} .$$

This statement is obviously true if all of the implied derivatives exist. Since $g_a$ is assumed to be of class $C^2$ and $y(z^k) \in \mathfrak{D}_1(D^*)$, the only terms that are in question are of the form

$$\frac{\partial^2 g_a}{\partial [\partial_n y(z)] \, \partial [\partial_m y(z)]} \partial_n \partial_m y(z) .$$

Thus, we must demand that $y(z^k)$ have *second derivatives* throughout $D^*$, with the possible exception of those points where

$$\frac{\partial^2 g_a}{\partial [\partial_n y(z)] \, \partial [\partial_m y(z)]} = 0 .$$

If these additional conditions are met, the appropriate continuity structure is available for the application of the divergence theorem. An application of the divergence theorem and the fact that $\delta y(z^k)$ vanishes on the boundary of $D^*$ gives

$$k_a(x^k; y + \delta y) = k_a(x^k; y) + \delta k_a(x^k; y) + o(\| \delta y \|) \tag{1.20}$$

where

$$\delta k_a(x^k; y) = \int_{D^*} \left\{ \frac{\partial g_a}{\partial y(z)} - \frac{\partial}{\partial z^m} \left[ \frac{\partial g_a}{\partial(\partial_m(z))} \right] \right\} \delta y(z)\, dV(z) . \tag{1.21}$$

Additional notation must now be introduced so that the content of the statements will not be obscured. From (1.15) we see that $x$ may be viewed as parameters and $k_a$ as $q$ functionals on $\mathfrak{D}_1(D^*)$ defined by the lagrangian functions $g_a$. Let us, therefore, introduce the operator

$$\left\{ e \mid g_a(x^k) \right\}_y (z^k) = \frac{\partial g_a}{\partial y(z)} - \frac{\partial}{\partial z^m} \left\{ \frac{\partial g_a}{\partial[\partial_m(z)]} \right\}, \tag{1.22}$$

in which the symbols have the following meanings: (1) $x$ within the parenthesis immediately after $g_a$ denote parameters which are held fixed in calculating the derivatives with respect to $y(z^k)$ and $\partial_m y(z^k)$, (2) $z$ denote coordinates of a point in $X_n$ at which the quantity on the right-hand side of (1.22) is evaluated, and (3) $y$ is a function of $z$. Having defined the above terms, we can write (1.21) in the equivalent form

$$\delta k_a(x^k; y) = \int_{D^*} \left\{ e \mid g_a(x^k) \right\}_y (z^k)\, \delta y(z^k)\, dV(z) \tag{1.23}$$

When the above results are substituted into (1.19) we obtain

$$\Delta J[y](L) = \int_{D^*} \left[ \frac{\partial L}{\partial y} \delta y + \frac{\partial L}{\partial(\partial_m y)} \partial_m \delta y + \frac{\partial L}{\partial k_a} \delta k_a \right] dV(x) + o(\| \delta y \|) . \tag{1.24}$$

Since $L$ is of class $C^2$, the divergence theorem and the fact that $\delta y$ vanishes on $D^*$ can be used to obtain

$$\int_{D^*} \left\{ \frac{\partial L}{\partial y} \delta y + \frac{\partial L}{\partial(\partial_m y)} \partial_m \delta y \right\} dV(x) = \int_{D^*} \left\{ e \mid L(k_a) \right\}_y (x^k)\, \delta y(x^k)\, dV(x) \tag{1.25}$$

with the provision that $y(x^k)$ possesses second derivatives at all points of $D^*$ with the possible exception of those points at which

$$\frac{\partial^2 L}{\partial(\partial_n y)\, \partial(\partial_m y)} = 0 .$$

If (1.23) is substituted into the last term on the right-hand side of (1.24), we have

$$\int_{D^*} \frac{\partial L}{\partial k_a} \delta k_a \, dV(x) = \int_{D^*}\int_{D^*} \frac{\partial L}{\partial k_a}(x^k) \Big\}e \,|\, g_a(x^k)\Big\}_y (z^k)\,\delta y(z^k)\,dV(z)\,dV(x) \tag{1.26}$$

where $(\partial L/\partial k_a)(x^k)$ denotes that the expression is a function of $x$ since all of its arguments are functions of $x$ alone.

The only complicated step in the computation is that the variation $\delta y(z^k)$ appears in (1.26). Since the domains of $x$ and $z$ are the same in (1.26), the result is the same if we interchange $x$ and $z$. If we use the symbol * to denote this interchange of $x$ and $z$, then $(AB)^* = A^*B^*$ , and

$$\left[\frac{\partial L}{\partial k_a}(x^k)\right]^* \equiv \frac{\partial L}{\partial k_a}(z^k)$$

$$\left[\Big\}e \,|\, g_a(x^k)\Big\}_y (z^k)\right]^* \equiv \Big\}e \,|\, g_a^*(z^k)\Big\}_y (x^k).$$

Here $g_a^*$ is defined by

$$\begin{aligned} g_a &= g_a\left[x^k, z^k, y(z^k), \partial_m y(z^k)\right] \\ g_a^* &= g_a\left[z^k, x^k, y(x^k), \partial_m y(x^k)\right]. \end{aligned} \tag{1.27}$$

A combination of (1.24), (1.25), and (1.27) then gives

$$\Delta J[y](L) =$$

$$\int_{D^*}\left[\Big\}e \,|\, L(k_a)\Big\}_y (x^k) + \int_{D^*} \frac{\partial L}{\partial k_a}(z)\Big\}e \,|\, g_a^*(z)\Big\}_y (x^k)\,dV(z)\right]\delta y(x^k)\,dV(x) + o(\|\delta y\|). \tag{1.28}$$

Hence the linear approximating functional $j[y;\delta y] = \delta J[y](L)$ is given by

$$\int_{D^*}\left[\Big\}e \,|\, L(k_a)\Big\}_y (x^k) + \int_{D^*} \frac{\partial L}{\partial k_a}(z)\Big\}e \,|\, g_a^*(z)\Big\}_y (x^k)\,dV(z)\right]\delta y(x^k)\,dV(x).$$

When Lemma 1.2 is used, the following theorem obtains.

**Theorem 1.2.** *Let* $J[y](L)$ *be defined by* (1.15) *and* (1.16) *in which* $L$ *and* $g_a$ *are functions of class* $C^2$ *of their indicated arguments. If* $y_0$ *is a function of class* $C^2$ *on* $D^*$, *then* $J[y](L)$ *is stationary at* $y = y_0$ *if and only if the Euler equation*

$$\{\mathcal{E}\,|\,L\,\}_{y_0}(x^k) = 0 \tag{1.29}$$

*holds in* $D^*$, *where*

$$\{\mathcal{E}\,|\,L\,\}_y(x^k) \equiv \Big\{e\,|\,L(k_a)\Big\}_y(x^k) + \int_{D^*}\frac{\partial L}{\partial k_a}(z^k)\Big\{e\,|\,g_a^*(z^k)\Big\}_y(x^k)\,dV(z) \tag{1.30}$$

*is the (nonlocal) Euler-Lagrange operator and*

$$\{e\,|\,W(v)\,\}_y(x^m) = \frac{\partial W}{\partial y} - \frac{\partial}{\partial x^m}\left[\frac{\partial W}{\partial(\partial_m y)}\right] \tag{1.31}$$

($v$ *denoting parameters that are held constant during the indicated differentiation with respect to* $y$ *and* $\partial_m y$) *is the little Euler-Lagrange operator.*

The restriction to $y_0 \in C^2$ results because of the necessity for integration by parts and application of the divergence theorem. In the case of only *one independent variable*, it can be shown that if $y_0$ stationarizes $J[y](L)$ and $y_0 \in C^1$, then $y_0$ has a second derivative where $\partial^2 L/\partial^2 y_0' \neq 0$. It does not appear possible to establish the above for several independent variables. However, we can relax the required continuity conditions on $y_0$ to some degree, since it was only required that $y_0$ have second derivatives whenever

$$\frac{\partial^2 L}{\partial(\partial_m y)\,\partial(\partial_m y)} \neq 0\,, \qquad \frac{\partial^2 g_a}{\partial(\partial_m y)\,\partial(\partial_n y)} \neq 0\,.$$

When these conditions are traced through the argument leading to (1.29) (in particular, the considerations involving interchange of $x$ and $z$) we have the following result.

**Theorem 1.3.** *Let* $J[y](L)$ *be defined by* (1.15) *and* (1.16) *in which* $L$ *and* $g_a$ *are functions of class* $C^2$ *of their indicated arguments. If* $y_0$ *has second derivatives at all points of* $D^*$ *except those for which*

$$\frac{\partial^2 L}{\partial(\partial_n y)\,\partial(\partial_m y)} + \int_{D^*} \frac{\partial L}{\partial k_a}(z) \frac{\partial^2 g_a^*}{\partial(\partial_n y)\,\partial(\partial_m y)} dV(z) = 0,$$

*then* $J[y](L)$ *is stationary at* $y = y_0$ *if and only if* $y_0$ *is a solution of the Euler equation* (1.29).

The meaning of the various operations required by (1.30) and (1.31) is best clarified by an example. Set

$$g_1\left[x^k, z^k, y(z^k), \partial_m y(z^k)\right] = K(x^k, z^k)\, y(z^k) + H^m(x^k, z^k)\, \partial_m y(z^k), \quad (132)$$

then

$$k_1(x^k; y) = \int_{D^*} \left[K(x^k, z^k)\, y(z^k) + H^m(x^k, z^k)\, \partial_m y(z^k)\right] dV(z). \quad (1.33)$$

Also set

$$L\left[x^k, y(x^k), \partial_m y(x^k), k_1(x^k; y)\right] = y(x^k)\left[f(x^k) + k_1(x^k; y) - \tfrac{1}{2} y(x^k)\right], \quad (1.34)$$

so that

$$J[y](L) = \int_{D^*} y(x^k)\left[f(x^k) + k_1(x^k; y) - \tfrac{1}{2} y(x^k)\right] dV(x). \quad (1.35)$$

From (1.31) we obtain

$$\left\{e \,|\, L(k_1)\right\}_y (x^k) = \left.\frac{\partial L}{\partial y}\right|_{k_1} - \partial_m \left[\left.\frac{\partial L}{\partial(\partial_m y)}\right|_{k_1}\right] = -y(x^k) + f(x^k) + k_1(x^k; y),$$

(1.36)

since $L(k_1)$ within the braces denotes that $k_1$ is to be held constant during the differentiation with respect to $y$ and $\partial_m y$. A combination of (1.27) and (1.32) gives

$$g_1^* = K(z^k, x^k)\, y(x^k) + H^m(z^k, x^k)\, \partial_m y(x^k),$$

and hence

$$\left\{e \,|\, g_1^*(z^k)\right\}_y (x^k) = K(z^k, x^k) - \frac{\partial H^m(z^k, x^k)}{\partial x^m}. \quad (1.37)$$

Now, $\partial L/\partial k_1 = y(x^k)$, and hence we obtain

$$\frac{\partial L}{\partial k_1}(z^k) = y(z^k).$$

Accordingly, we have

$$\int_{D^*} \frac{\partial L}{\partial k_1}(z^k)\left\{e \mid g_a^*(z^k)\right\}_y (x^k)\, dV(z) = \int_{D^*} y(z^k)\left[K(z^k, x^k) - \frac{\partial H^m(z^k, x^k)}{\partial x^m}\right] dV(z). \tag{1.38}$$

By substituting the above results into (1.30) we finally obtain

$$\{\mathcal{E} \mid L\}_y (x^k) = -y(x^k) + f(x^k) + \int_{D^*} [K(x^k; z^k) + K(z^k, x^k)]\, y(z^k)\, dV(z)$$
$$+ \int_{D^*}\left[H^m(x^k, z^k)\frac{\partial y(z^k)}{\partial z^m} - y(z^k)\frac{\partial H^m(z^k, x^k)}{\partial x^m}\right] dV(z). \tag{1.39}$$

Noting that $L$ is independent of the derivatives and that $g_1$ is a linear function of the derivatives, we see that the second derivatives of $L$ and of $g_1$ with respect to the derivatives of $y(x^k)$ vanish throughout $D^*$. We may consequently apply Theorems 1.1 - 1.3 and conclude that $y(x^k) \in S(L)$ if and only if

$$y(x^k) = f(x^k) + \int_{D^*} \{K(x^k, z^k) + K(z^k, x^k)\}\, y(z^k)\, dV(z)$$
$$+ \int_{D^*}\left[H^m(x^k, z^k)\frac{\partial y(z^k)}{\partial z^m} - y(z^k)\frac{\partial H^m(z^k, x^k)}{\partial x^m}\right] dV(z). \tag{1.40}$$

## 1.6. PROPERTIES OF THE EULER-LAGRANGE OPERATOR

Let $L_1$ and $L_2$ be two lagrangian functions, and let $a$ and $b$ be two real numbers. The following results are established from the defining equations (1.30) and (1.31):

$$\left\{\mathcal{E} \mid aL_1 + bL_2\right\}_y (x^k) = a\left\{\mathcal{E} \mid L_1\right\}_y (x^k) + b\left\{\mathcal{E} \mid L_2\right\}_y (x^k), \tag{1.41}$$

$$\left\{\mathcal{E} \mid L_1 L_2\right\}_y (x^k) \neq L_1\left\{\mathcal{E} \mid L_2\right\}_y (x^k) + L_2\left\{\mathcal{E} \mid L_1\right\}_y (x^k). \tag{1.42}$$

Hence, *the Euler-Lagrange operator is a linear operator on the class of all lagrangian functions*, but it is not a derivate operator (it does not satisfy Leibniz's rule). This linearity property will be of repeated use.

If the lagrangian function is given, application of the Euler-Lagrange operator will, in general, result in an integrodifferential operator acting on $y(x^k)$. This resulting integrodifferential operator on $y(x^k)$ is usually well defined only for elements $y(x^k)$ taken from a part of $\mathfrak{D}_1(D^*)$. If the lagrangian function is a positive definite qudratic form in the first partial derivatives of $y(x^k)$, then it must be assumed that $y(x^k)$ is of class $C^2$ on $D^*$. This assumption is based on the fact that the Euler-Lagrange operator applied to such a lagrangian function will involve the second derivatives of $y(x^k)$ with nonvanishing coefficients throughout $D^*$. For each given lagrangian function $L$, the collection of all functions on $D^*$ for which every term of $\{\mathcal{E}|L\}_y(x^k)$ is well defined will be referred to as the $L$*-domain* and will be denoted by $\#L$. For instance, if $L$ involves only linear functions of the first derivatives of $y(x^k)$, then $\#L$ coincides with $\mathfrak{D}_1(D^*)$; if $L$ is independent of the first derivatives of $y(x^k)$, then $\#L$ coincides with $\mathcal{C}(D^*)$; and if $L$ contains a positive definite quadratic form in the first derivatives of $y(x^k)$, then $\#L$ is the subclass of $\mathfrak{D}_1(D^*)$ comprised of all $C^2$ functions on $D^*$.

If

$$\{\mathcal{E}|L_1\}_y(x^k) = \{\mathcal{E}|L_2\}_y(x^k) \tag{1.43}$$

holds for all $y(x^k)$ contained in $\#L_1 \cap \#L_2$, what can be concluded concerning the relationship between $L_1$ and $L_2$? From the linearity of the Euler-Lagrange operator, (1.43) can hold if and only if

$$\{\mathcal{E}|L_1 - L_2\}_y(x^k) = 0 \tag{1.44}$$

for all $y(x^k)$ in $\#L_1 \cap \#L_2$. Let $\mathfrak{N}$ denote the collection of all lagrangian functions for which

$$\{\mathcal{E}|S\}_y(x^k) = 0 \tag{1.45}$$

for all $y(x^k)$ in $\mathfrak{D}_1(D^*)$ (that is, $\#S = \mathfrak{D}_1(D^*)$). The collection $\mathfrak{N}$ is referred to as the *null class of the Euler-Lagrange operator*. The following result is now a direct consequence of (1.44) and the definition of $\mathfrak{N}$. *If* $L_1$ *and* $L_2$ *are two lagrangian functions such that*

$$\{\mathcal{E}|L_1\}_y(x^k) = \{\mathcal{E}|L_2\}_y(x^k) \tag{1.46}$$

*holds for all* $y(x^k)$ *in* $\#L_1 \cap \#L_2$, *then there exists an element* $S$ *of* $\mathfrak{N}$ *such that*

$$L_1 = L_2 + S \tag{1.47}$$

*holds for all* $y(x^k)$ *in* $\mathfrak{D}_1(D^*)$ *and for all* $x^k$ *in* $D^*$.

Since the sum of any two elements of $\mathfrak{N}$ is an element of $\mathfrak{N}$, as follows from the linearity of the Euler-Lagrange operator, the relation

$$L_1 \mathrel{\hat{=}} L_2 \Leftrightarrow L_1 = L_2 + S, \qquad S \in \mathfrak{N} \tag{1.48}$$

is obviously symmetric, reflective, and transitive, and hence is an equivalence relation on the collection of all lagrangian functions. This equivalence relation can be used to partition the collection of all lagrangian functions into disjoint equivalence classes. We thus have, if $y_0(x^k) \in S(L)$, then $y_0(x^k) \in S(\hat{L})$, where $\hat{L}$ denotes *the equivalence class of lagrangian functions containing* L.

An equivalent way of characterizing $\mathfrak{N}$ follows directly from the defining equation (1.45): *the null class of the Euler-Lagrange operator consists of all lagrangian functions* L *for which every element of* $\mathfrak{D}_1(D^*)$ *is a stationarization element of* $J[y](L)$.

We obviously require a detailed representation of the lagrangian functions that comprise $\mathfrak{N}$. First, we examine those functionals which correspond to functions in the ordinary calculus having identically vanishing derivatives. If the lagrangian function $L$ is such that $J[y](L)$ is constant in value for all elements of $\mathfrak{D}_1(D^*)$, the variation of $J[y](L)$ vanishes for all variations of $y(x^k)$, in contrast to vanishing only for those variations of $y(x^k)$ which vanish on $\partial D^*$. Such lagrangian functions belong to $\mathfrak{N}$ since the linear approximating functional vanishes identically for all elements of $\mathfrak{D}_1(D^*)$. *The trivial subclass*, $\mathfrak{N}^*$, *consists of all lagrangian functions for which* $J[y](L)$ *is a constant valued functional on* $\mathfrak{D}_1(D^*)$.

As an immediate example, consider the lagrangian function

$$L = y(x^k)H^m(x^k)\partial_m y(x^k) - \frac{1}{2}H^m(x^k)\partial_m\left[y(x^k)^2\right] + u(x^k).$$

We than have $J[y](L) = \int_{D^*} u(x^k)\,dV(x)$, and hence $L \in \mathfrak{N}^*$. However, some elements of $\mathfrak{N}^*$ are not quite so obvious. For instance, consider the lagrangian function

$$L = y(x^k)\int_{D^*} H^m(z^k)\partial_m y(z^k)\,dV(z) - H^m(x^k)\partial_m y(x^k)\int_{D^*} y(z^k)\,dV(z).$$

In this case we have

$$J[y](L) = \int_{D^*}\int_{D^*}\left[y(x^k)H^m(z^k)\partial_m y(z^k) - y(z^k)H^m(x^k)\partial_m y(x^k)\right] dV(z)\,dV(x).$$

Since the domains of the two integrals are the same, we can interchange the order of integration in the second term and obtain

$$J[y](L) = \int_{D^*}\int_{D^*}\left[y(x^k)H^m(z^k)\partial_m y(z^k) - y(x^k)H^m(z^k)\partial_m y(z^k)\right] dV(z)\,dV(x) = 0.$$

Hence $L$ belongs to $\mathfrak{N}^*$.

In the ordinary calculus only the constant functions have everywhere vanishing derivatives. The calculus of functionals, however, has a much wider class of functionals with everywhere vanishing variation. This wider class of $\mathfrak{N}$ results because only variations of $y(x^k)$ which vanish on the boundary of $D^*$ are considered. This can be seen by considering the lagrangian function

$$L = \partial_m f^m[x^k, y(x^k)],$$

in which the $n$ functions $f^1, \ldots, f^n$ are of class $C^3$ in their arguments (that is, $L$ is of class $C^2$ in its arguments). We then have

$$J[y](L) = \int_{D^*}\partial_m f^m[x^k, y(x^k)]\,dV(x) = \int_{\partial D^*} f^m[x^k, y(x^k)]\,dS_m(x),$$

and hence

$$J[y+\delta y](L) - J[y](L) = \int_{\partial D^*}\frac{\partial f^m[x^k, y(x^k)]}{\partial y(x^k)}\,\delta y(x^k)\,dS_m(x) + o(\|\delta y\|).$$

Thus, since $\delta y(x^k)$ vanishes on the boundary of $D^*$, the linear approximating functional vanishes identically; and hence, $L \in \mathfrak{N}$.

At this point, we introduce a notational convenience. Let $U[x^k, y(x^k), v(x^k)]$ be a function of class $C^1$ in its indicated arguments. We write

$$\partial_m U[x^k, y(x^k), v(x^k)] = \partial^*_m U[x^k, y(x^k), v(x^k)] + \frac{\partial U(x^k, y(x^k), v(x^k)}{\partial y(x^k)}\,\partial_m y(x^k), \tag{1.49}$$

where $\partial^*_m U$ denotes the partial derivative of $U$ with respect to $x^k$ for constant $y(x^k)$.

**Lemma 1.6.** *If* L *is such that*

$$\{e|L(v)\}_y(x^k) = 0 \tag{1.50}$$

*for all* $C^2$ *functions* $y(x^k)$ *on* $D^*$, *where* $v(x^k)$ *is any quantity that is held constant during the indicated differentiation with respect to* $y(x^k)$ *and its derivatives, then* $\#L = \mathfrak{D}_1(D^*)$. *The most general* L *for which* (1.50) *holds for all* $y(x^k) \in \mathfrak{D}_1(D^*)$ *is given by*

$$L = U[x^k, y(x^k), v(x^k)] + U^m[x^k, y(x^k), v(x^k)]\,\partial_m y(x^k), \tag{1.51}$$

*where the* $n+1$ *functions* $U[x^k, y(x^k), v(x^k)]$ *and* $U^m[x^k, y(x^k), v(x^k)]$ *are of class* $C^2$ *and stand in the relation*

$$\partial^*_m U^m = \frac{\partial U}{\partial y}. \tag{1.52}$$

**Proof.** When (1.31) is used to expand the left-hand side of (1.50), we have

$$\frac{\partial L}{\partial y} - \partial^*_m\left[\frac{\partial L}{\partial(\partial_m y)}\right] - \left[\frac{\partial^2 L}{\partial y\,\partial(\partial_m y)}\right]\partial_m y - \left[\frac{\partial^2 L}{\partial(\partial_n y)\,\partial(\partial_m y)}\right]\partial_m\partial_n y = 0. \tag{1.53}$$

If (1.53) is to hold for all $C^2$ functions on $D^*$, the coefficients of the first and second derivatives of $y(x^k)$ must vanish identically. Accordingly, we must have

$$\frac{\partial^2 L}{\partial(\partial_n y)\,\partial(\partial_m y)} = 0,$$

and hence $L$ must have the form given by (1.51). This form, however, is that which is required in order that $\#L = \mathfrak{D}_1(D^*)$: the little Euler-Lagrange operator applied to a lagrangian function which is linear in the derivatives of $y(x^k)$ does not require second derivatives of $y(x^k)$. We may thus consider the satisfaction of (1.50) for all elements of $\mathfrak{D}_1(D^*)$. Substituting (1.51) into (1.53) gives

$$\frac{\partial U}{\partial y} + \frac{\partial U^m}{\partial y}\partial_m y - \partial^*_m U^m - \frac{\partial U^m}{\partial y}\partial_m y = \frac{\partial U}{\partial y} - \partial^*_m U^m = 0,$$

and, thus, (1.52) is obtained. The continuity requirements of $U$ and $U^m$ follow from the fact that all lagrangian functions are assumed to be of class $C^2$.

**Lemma 1.7.** *A $C^2$ function $L(x^k, y(x^k), v(x^k))$ is such that*

$$\{e \mid L\}_y(x^k) = 0 \tag{1.54}$$

*for all $y(x^k) \in \mathfrak{D}_1(D^*)$ if and only if there exist $n$ functions $V^m[x^k, y(x^k), v(x^k)]$ of class $C^3$ in their indicated arguments such that*

$$L = \partial_m V^m[x^k, y(x^k), v(x^k)]. \tag{1.55}$$

*The quantities V are related to the U in (1.51) by*

$$V^m[x^k, y(x^k)] = \int_0^{y(x^k)} U^m[x^k, t, v(x^k)]\, dt; \tag{1.56}$$

*that is,*

$$U^m = \frac{\partial V^m}{\partial y}, \qquad U = \partial^*_m V^m. \tag{1.57}$$

**Proof.** The relations (1.52) are the integrability conditions for the system (1.57). Thus, since $(U, U^m)$ are $C^2$ functions, the complete integrability of the system (1.57) yields the solution (1.56), so that $V$ exist. A substitution of (1.57) into (1.51) and use of Lemma 1.7 gives (1.55). Further, $V$ are of class $C^3$, for if $(U, U^m)$ are of class $C^2$ and satisfy the relations (1.52), then each $U^m$ must be of class $C^3$ in $x$ and of class $C^2$ in $y(x^k)$. The integration in (1.56) gives the $C^3$ structure.

Lemmas 1.8 and 1.9 will be proved simultaneously.

**Lemma 1.8.** *If L is such that*

$$\{\mathcal{E} \mid L\}_y(x^k) = 0 \tag{1.58}$$

*for all $C^2$ functions $y(x^k)$ on $D^*$, then $\#L = \mathfrak{D}_1(D^*)$.*

**Lemma 1.9** *We have*

$$\{\mathcal{E} \mid L\}_y(x^k) = \left\{ e \mid L + \int_{D^*} \frac{\partial L}{\partial k_a}(z)\, g^*_a\, dV(z) \right\}_y (x^k) \tag{1.59}$$

*for all $y(x^k) \in \#L$.*

**Proof.** If we write $\{\mathcal{E}|L\}_y(x^k)$ in full, we have

$$\{\mathcal{E}|L\}_y(x^k) = \frac{\partial L}{\partial y} - \partial_m^*\left[\frac{\partial L}{\partial(\partial_m y)}\right] - \left[\frac{\partial^2 L}{\partial y\,\partial(\partial_m y)}\right]\partial_m y - \left[\frac{\partial^2 L}{\partial(\partial_n y)\partial(\partial_m y)}\right]\partial_m\partial_n y$$

$$+ \int_{D^*}\frac{\partial L}{\partial k_a}(z)\left\{\frac{\partial g_a^*}{\partial y} - \partial_m^*\left[\frac{\partial g_a^*}{\partial(\partial_m y)}\right] - \left[\frac{\partial^2 g_a^*}{\partial y\,\partial(\partial_m y)}\right]\partial_m y\right. \qquad (1.60)$$

$$\left. - \left[\frac{\partial^2 g_a^*}{\partial(\partial_n y)\,\partial(\partial_m y)}\right]\partial_m\partial_n y\right\} dV(z)$$

where $(\partial L/\partial k_a)(z)$ depends only on $z^k$ and functions of $z^k$ and all the derivatives within the braces under the integral sign are with respect to $x^k$ or with respect to $y(x^k)$ and its derivative. If (1.60) is to vanish for all $y(x^k)$ that are of class $C^2$ on $D^*$, then the coefficients of $y(x^k)$ and its derivatives must vanish separately. Equating the coefficients of the second derivatives of $y(x^k)$ to zero gives

$$\frac{\partial^2 L}{\partial(\partial_n y)\,\partial(\partial_m y)} + \int_{D^*}\frac{\partial L}{\partial k_a}(z)\,\frac{\partial^2 g_a^*}{\partial(\partial_n y)\,\partial(\partial_m y)}\,dV(z) = 0, \qquad (1.61)$$

since the second partial derivatives of $y(x^k)$ can be taken outside of the integral with respect to $z$. The conditions (1.61) are those under which we can relax the requirement that $y(x^k) \in C^2$ This was shown in Sec. 1.5 in the proof of the basic theorem on stationarization elements. We thus have that $\#L = \mathfrak{D}_1(D^*)$. In view of the assumed continuity structure of $L$ and $g_a$, the differentiations with respect to $x^m$, $y(x^m)$ and $\partial_m y(x^m)$ and the $z$-wise integration appearing in (1.60) can be commuted. If this is done, we have (1.59).

A direct combination of the above results leads to the following conclusion.

**Theorem 1.4.** *The most general element of the null class of the Euler-Lagrange operator is either an element of* $\mathfrak{N}^*$ *or it is such there exists a collection of functionals*

$$K_p(x^k; y) = \int_{D^*} G_p\left[x^k, z^k, y(z^k), \partial_m y(z^k)\right] dV(z)$$

*and a collection of* $n$ *functions* $V^m\left(x^k, y(x^k), K_p(x^k; y)\right)$ *of class* $C^3$ *in the variables* $x^k$ *and* $y(x^k)$ *such that*

$$L + \int_{D^*} \frac{\partial L}{\partial k_a}(z)\, g_a^*\, dV(z) = \partial_m V^m\left[x^k, y(x^k), K_p(x^k; y)\right]. \qquad (1.62)$$

The following example demonstrates that $K_p$ cannot, in general, be identified with $k_a$. In the case in which there is only one $K_p$, we can write simply $K$. We then have

$$\partial_m V^m[x^k, y(x^k), K(x^k; y)] = \frac{\partial V^m}{\partial x^m} + \frac{\partial V^m}{\partial y}\partial_m y + \frac{\partial V^m}{\partial K}\partial_m K.$$

Accordingly, $\partial_m V^m$ involves not just the functional $K$, but rather the $n+1$ functionals $K$ and $\partial_m K$. Thus, if (1.62) is to hold, $k_a$ must be identified in some fashion with the $n+1$ functionals $K$ and $\partial_m K$.

Equation (1.62) is, in general, an extremely complicated non-linear, functional substitution, integral equation for the determination of $L$. It cannot be solved directly and, thus, an alternative path must be sought. First, it is shown that (1.62) is satisfied by a divergence whose form is similar to the right-hand side of (1.62), with the added condition that the functions $G_p$ which give rise to $K_p$ are themselves divergences. This step is reasonably straightforward, and the answer is as expected, for such a lagrangian function leads to a functional $J[y](L)$ which is determined solely from the values which $y(x^k)$ assumes on the boundary of $D^*$ (remember that the variation of $y(x^k)$ vanishes on the boundary of $D^*$ in computing the Euler-Lagrange operator as the quantity determining the value of $\delta J$). Second, it is shown that any lagrangian function that cannot be represented as such a divergence is a member of $\mathfrak{N}$ if and only if the functional that it gives rise to is a constant valued function and is an element of $\mathfrak{N}^*$. To demonstrate this would require considerably more functional analysis than is assumed for this presentation. Therefore, we shall simply state the final result.

**Theorem 1.5.** *The most general element of the null class of the Euler-Lagrange operator is of the form*

$$A + \partial_m V^m\left[x^k, y(x^k), K_p(x^k; y)\right],$$

*in which* $A$ *is an element of* $\mathfrak{N}^*$,

$$K_p(x^k; y) = \int_{D^*} G_p\left[x^k, z^k, y(z^k), \partial_m y(z^k)\right] dV(z)$$

*and*

$$G_p = \frac{\partial}{\partial z^m} f_p{}^m[x^k, z^k, y(z^k)].$$

If we recall that

$$\{\mathcal{E}|L\}_y(x^k) = \}e|L(k_a)\{_y(x^k) + \int_{D^*} \frac{\partial L}{\partial k_a}(z)\, \}e|g_a^*(z^k)\{_y(x^k)\, dV(z),$$

then Theorem 1.5 says in effect that we first annihilate $\}e|g_a^*(z^k)\{_y(x^k)$ for every $y(x^k) \in \mathfrak{D}_1(D^*)$, and then annihilate $\}e|L(k_a)\{_y(x^k)$ for every $y(x^k) \in \mathfrak{D}_1(D^*)$. If $z$ had not been interchanged with $x$ to obtain the Euler-Lagrange operator, this would be equivalent to the statement that $\delta J = 0$ for all $y(x^k) \in \mathfrak{D}_1(D^*)$, if and only if the coefficients of $y(x^k)$, $\partial_m y(x^k)$, $\partial_m \partial_n y(x^k)$, and the integrals of these quantities must vanish separately.

The statement that "a divergence is variationally deletable," is often encountered in the literature. From Theorem 1.5, it is shown that this is only the case when $V^m$ and $f_p{}^m$ are not explicit functions of the derivatives of $y(x^k)$. To avoid errors it is important not to overlook this added condition which arises from the fact that the variation of $y(x^k)$ is assumed to vanish on the boundary of $D^*$ while no assumptions are made concerning the values of the derivative of the variation on the boundary of $D^*$. From another point of view, we have assumed that $L$ and $k_a$ involve no derivatives higher than the first, while a divergence of a vector valued function of first derivatives would involve second derivatives. If we were to consider lagrangian functions whose arguments are second derivatives as well, we could then entertain first derivative arguments in $V^m$ and $f_p{}^m$. However, to do so would necessitate the use of the space of $C^2$ functions with the uniform convergence norm $\mathfrak{D}_2(D^*)$. In this case it also follows that if $\{\mathcal{E}|L\}_y(x^k) = 0$ holds for all $C^4$ functions $y(x^k)$, then $\#L = \mathfrak{D}_2(D^*)$. The proof is the same but lengthier than that given above.

At first glance, the following lagrangian function might not appear to belong to $\mathfrak{N}$:

$$L = y(x^k)\int_{D^*} \frac{\partial H^m(z^k, x^k)}{\partial x^m}\, dV(z) + \int_{D^*} H^m(x^k, z^k)\, \partial_m y(z^k)\, dV(z).$$

However, a little manipulation gives the relation

$$L = \frac{\partial}{\partial x^m}\left[y(x^k)\int_{D^*} H^m(z^k, x^k)\,dV(z)\right]$$

$$+ \int_{D^*} [H^m(x^k, z^k)\,\partial_m y(z^k) - H^m(z^k, x^k)\,\partial_m y(x^k)]\,dV(z)\,.$$

Hence, $L$ indeed belongs to $\mathfrak{N}$ since the first term is a divergence and the second term is an element of $\mathfrak{N}^*$.

## 1.7. LAGRANGIANS WHICH RESULT IN LINEAR OPERATORS

An operator $\mathcal{O}$ is said to be a linear operator if and only if

$$\mathcal{O}(a\cdot y_1 + b\cdot y_2) = a\mathcal{O}(y_1) + b\mathcal{O}(y_2)$$

for any numbers $a$ and $b$ and any two quantities $y_1$ and $y_2$ which are elements of a linear space and lie in the domain of $\mathcal{O}$. It is evident from the definition of the Euler-Lagrange operator (see (1.30) and (1.31)) that *this operator is linear if and only if the lagrangian function is a quadratic form in* $y$ *and its first partial derivatives. The Euler equation arising from a lagrangian function is a linear equation if and only if the lagrangian function is a polynomial function of* $y$ *and its derivatives of degree less than three. Also, the Euler equation is linear and homogeneous if and only if the lagrangian function is a polynomial function of* $y$ *and its derivatives of degree less than three without linear terms.* Since the Euler-Lagrange operator is known to be linear with respect to lagrangian functions themselves, we can characterize all lagrangian functions that result in linear Euler equations by studying each possible term that makes up the polynomial lagrangian.

We define the quantities $g_a$, $a = 1, \ldots, 5$ by

$$\begin{aligned}
g_1 &= K(x^k, z^k)\,y(z^k)\,, & g_1^* &= K(z^k, x^k)\,y(x^k)\,,\\
g_2 &= H^m(x^k, z^k)\,\partial_m y(z^k)\,, & g_2^* &= H^m(z^k, x^k)\,\partial_m y(x^k)\,,\\
g_3 &= k(x^k, z^k)\,y(z^k)^2, & g_3^* &= k(z^k, x^k)\,y(x^k)^2,\\
g_4 &= h^m(x^k, z^k)\,y(z^k)\,\partial_m y(z^k)\,, & g_4^* &= h^m(z^k, x^k)\,y(x^k)\,\partial_m y(x^k)\,,\\
g_5 &= H^{nm}(x^k, z^k)\,\partial_n y(z^k)\,\partial_m y(z^k)\,, & g_5^* &= H^{nm}(z^k, x^k)\,\partial_n y(x^k)\,\partial_m y(x^k)\,.
\end{aligned} \tag{1.63}$$

We define $k_a(x^k; y)$ by

$$k_a(x^k; y) = \int_{D^*} g_a\left[x^k, z^k, y(z^k), \partial_m y(z^k)\right] dV(z). \tag{1.64}$$

The following list spans the set of all possible polynomial lagrangian functions of degree less than three, in which the $a$'s are assumed to be functions of $x$:

1) $ay$
2) $a^m \partial_m y$
3) $ak_1$
4) $ak_2$
5) $ay^2$
6) $a^{mn} \partial_m y \partial_n y$ with $a^{mn} = a^{nm}$
7) $a^m y \partial_m y$
8) $ak_1^2$
9) $ak_2^2$
10) $ayk_1$
11) $ayk_2$
12) $a^m k_1 \partial_m y$
13) $a^m k_2 \partial_m y$
14) $ak_3$
15) $ak_4$
16) $ak_5$
17) $ak_1 k_2$
18) $\partial_m y \int_{D^*} H^m(x^k, z^k)\, y(z^k)\, dV(z)$.

The notation involved in Case 18 is somewhat cumbersome, and it would be advantageous to show that the Euler equation resulting from this case is the same as that resulting from Case 11. This equivalence is almost immediate, for

$$\partial_m y \int_{D^*} H^m(x^k, z^k)\, y(z^k)\, dV(z) - y \int_{D^*} H^m(z^k, x^k)\, \partial_m y(z^k)\, dV(z)$$

is an element of $\mathfrak{N}^*$. Hence, if $y(x^k)$ satisfies the Euler equation which results from Case 11 with $H^m(z^k, x^k)$ rather than $H^m(x^k, z^k)$, then $y(x^k)$ satisfies the Euler equation for Case 18, and conversely.

Listed below are the results of the application of the Euler-Lagrange operator on each of the above cases. Case 18 is not included since it is equivalent to case 11. The student is advised to work out each of the cases in detail.

1. $\{\mathcal{E} \mid ay\}_y(x^m) = a(x^m)$. (1.65)

2. $$\left\{\mathcal{E} \mid a^k \partial_k y\right\}_y (x^m) = -\partial_k a^k (x^m), \tag{1.66}$$

3. $$\left\{\mathcal{E} \mid ak_2\right\}_y (x^m) = \int_{D*} a(z^m) K(z^m, x^m)\, dV(z). \tag{1.67}$$

4. $$\left\{\mathcal{E} \mid ak_2\right\}_y (x^m) = -\int_{D*} a(z^m) \frac{\partial H^k(z^m, x^m)}{\partial x^k} dV(z), \tag{1.68}$$

5. $$\{\mathcal{E} \mid ay^2\}_y (x^m) = 2a(x^m)\, y(x^m). \tag{1.69}$$

6. $$\left\{\mathcal{E} \mid a^{km} \partial_k y \partial_m y\right\}_y (x^m) = -2 \frac{\partial \left[a^{km}(x^r)\, \partial_m y(x^r)\right]}{\partial x^k}, \qquad a^{km} = a^{mk}, \tag{1.70}$$

7. $$\left\{\mathcal{E} \mid a^k y\, \partial_k y\right\}_y (x^m) = -y(x^m) \frac{\partial a^k(x^m)}{\partial x^k}, \tag{1.71}$$

8. $$\left\{\mathcal{E} \mid ak_1^2\right\}_y (x^m) = \int_{D*} \hat{K}(\omega^m, x^m)\, y(\omega^m)\, dV(\omega), \tag{1.72}$$

where

$$K(\omega^m, x^m) = 2 \int_{D*} a(z^m) K(z^m, \omega^m) K(z^m, x^m)\, dV(z), \tag{1.73}$$

and hence

$$K(\omega^m, x^m) = K(x^m, \omega^m). \tag{1.74}$$

9. $$\left\{\mathcal{E} \mid ak_2^2\right\}_y (x^m) = \int_{D*} \hat{H}^k(\omega^m, x^m) \frac{\partial y(\omega^m)}{\partial \omega^k} dV(\omega), \tag{1.75}$$

where

$$\hat{H}^k(\omega^m, x^m) = -2 \int_{D*} a(z^m) \frac{\partial H^t(z^m, x^m)}{\partial x^t} H^k(z^m, \omega^m)\, dV(z), \tag{1.76}$$

and hence

$$\frac{\partial \hat{H}^k(\omega^m, x^m)}{\partial \omega^k} = \frac{\partial \hat{H}^k(x^m, \omega^m)}{\partial x^m}. \tag{1.77}$$

10. $$\{\mathcal{E} \mid ayk_1\}_y(x^m) = \int_{D^*} \overline{K}(x^m, z^m)\, y(z^m)\, dV(z), \tag{1.78}$$

where

$$\overline{K}(x^m, z^m) = a(x^m) K(x^m, z^m) + a(z^m) K(z^m, x^m), \tag{1.79}$$

and hence

$$\overline{K}(x^m, z^m) = \overline{K}(z^m, x^m). \tag{1.80}$$

11. $$\{\mathcal{E} \mid ayk_2\}_y(x^m) = \int_{D^*} \left[ \overline{H}^k(x^m, z^m) \frac{\partial y(z^m)}{\partial z^k} + \overline{K}(x^m, z^m)\, y(z^m) \right] dV(z), \tag{1.81}$$

where

$$\begin{aligned} \overline{H}^k(x^m, z^m) &= a(x^m) H^k(x^m, z^m) \\ \overline{K}(x^m, z^m) &= -a(z^m) \frac{\partial H^k(z^m, x^m)}{\partial x^k}, \end{aligned} \tag{1.82}$$

and hence

$$\frac{\partial \overline{H}^k(z^m, x^m)}{\partial x^k} = -\overline{K}(x^m, z^m). \tag{1.83}$$

12. $$\{\mathcal{E} \mid a^k k_1 \partial_k y\}_y(x^m) = \int_{D^*} \left[ \overline{H}^k(x^m, z^m) \frac{\partial y(z^m)}{\partial z^k} + \overline{K}(x^m, z^m)\, y(z^m) \right] dV(z), \tag{1.84}$$

where

$$\begin{aligned} \overline{H}^k(x^m, z^m) &= a^k(z^m) K(z^m, x^m) \\ \overline{K}(x^m, z^m) &= -\frac{\partial [a^k(x^m) K(x^m, z^m)]}{\partial x^k}, \end{aligned} \tag{1.85}$$

and hence

$$\frac{\partial \overline{H}(z^m, x^m)}{\partial x^k} = -\overline{K}(x^m, z^m) . \tag{1.86}$$

13. $$\{\mathcal{E} \mid a^k k_2 \partial_k y\}_y (x^m) = \int_{D^*} \overline{H}^k (x^m, z^m) \frac{\partial y(z^m)}{\partial z^k} dV(z) , \tag{1.87}$$

where

$$\overline{H}^k (x^m, z^m) = -\frac{\partial [a^r(x^m) H^k(x^m, z^m) + a^k(z^m) H^r(z^m, x^m)]}{\partial x^r} , \tag{1.88}$$

and hence

$$\frac{\partial \overline{H}^k (x^m, z^m)}{\partial z^k} = \frac{\partial \overline{H}^k (z^m, x^m)}{\partial x^k} . \tag{1.89}$$

14. $$\{\mathcal{E} \mid ak_3\}_y (x^m) = \hat{K}(x^m) y(x^m) , \tag{1.90}$$

where

$$\hat{K}(x^m) = 2 \int_{D^*} a(z^m) k(z^m, x^m) dV(z) . \tag{1.91}$$

15. $$\{\mathcal{E} \mid ak_4\}_y (x^m) = \hat{h}(x^m) y(x^m) , \tag{1.92}$$

where

$$\hat{h}(x^m) = -\int_{D^*} a(z^m) \frac{\partial h^k(z^m, x^m)}{\partial x^k} dV(z) . \tag{1.93}$$

16. $$\{\mathcal{E} \mid ak_5\}_y (x^m) = \frac{\partial \left[\hat{H}^{kr}(x^m) \partial_r y(x^m)\right]}{\partial x^k} , \tag{1.94}$$

where

$$\hat{H}^{kr}(x^m) = -\int_{D^*} 2a(z^m) H^{kr}(z^m, x^m) dV(z) , \qquad H^{kr} = H^{rk} . \tag{1.95}$$

$$17.\left\{\mathcal{E}\mid ak_1k_2\right\}_y(x^m) = \int_{D^*} \hat{H}^m(x^k,\tau^k)\,\partial_m y(\tau^k)\,dV(\tau) - \int_{D^*} \hat{K}(x^k,\tau^k)\,y(\tau^k)\,dV(\tau)\,,$$

where

$$H^m(x^k,\tau^k) = \int_{D^*} a(z^k)\,K(z^k,x^k)\,H^m(z^k,\tau^k)\,dV(z)\,,$$

$$\hat{K}(x^k,\tau^k) = \int_{D^*} a(z^k)\,\frac{\partial H^m(z^k,x^k)}{\partial x^m}\,K(z^k,\tau^k)\,dV(z)\,.$$

There are two reasons for listing the above results. First, certain similarities are immediately evident. Second, by appropriate choices of $a$ and the functions of $x$ and $z$ occurring in (1.63), the Euler equations for some of the 18 cases can be obtained by linear combinations of the remaining 18 cases. Careful inspection in this regard leads to the following result.

**Theorem 1.6.** *Any Lagrangian function that results in a linear Euler equation is a linear combination of Cases* 1, 5, 6, 8 *and* 9 *with appropriate choices of* $a(x^k), K(x^k,z^k)$ and $H^m(x^k,z^k)$. *That is*

$$L = a_1 y + a_2 y^2 + a^{mn}\partial_m y \partial_n y + a_3 {k_1}^2 + a_4 {k_2}^2. \tag{1.96}$$

*If the Euler equation is linear and homogeneous, so that the Euler-Lagrange operator applied to* L *is a linear operator on* $y(x^k)$, *we have*

$$L = a_2 y^2 + a^{mn}\partial_m y \partial_n y + a_3 {k_1}^2 + a_4 {k_2}^2\,. \tag{1.97}$$

All 18 cases were listed because it is often more convenient to handle a problem formulated in terms other than those of Cases 1, 5, 6, 8 and 9. The reader should now realize the scope and generality of the equations that can be handled from the structure obtained here. We can deal equally well with differential, integral, integro-differential and differential integrodifferential equations. As a consequence of theorem 1.6, it is important to note that any linear, homogeous equation (obtainable from a variational statement involving only first derivatives) can always be obtained from a lagrangian function that is a quadratic form in $y(x^k)$, the derivatives of $y(x^k)$, and the integrals of these quantities. It is also important to note that there are only nondiagonal terms of the form $a^{mn}\partial_m y \partial_n y$.

Let $\mathcal{O}$ be a linear operator on a class $\mathcal{R}$ of functions defined on $D^*$. A linear operator $\mathcal{O}^+$ on a class of functions $\mathcal{R}^+$ is said to be the *adjoint* (sometimes referred to as the formal adjoint) of the operator $\mathcal{O}$ if and only if

$$\int_{D^*} \{ w\mathcal{O}y - y\mathcal{O}^+w \}(x^k)\, dV(x^k) = \int_{\partial D^*} P^m(x^k; y; w)\, dS_m(x) \tag{1.98}$$

holds for all $y(x^k) \in \mathcal{R}$, for all $w(x^k) \in \mathcal{R}^+$, where $P^m(x^k; y; w)$ is the same system of $n$ operators for all $y(x^k) \in \mathcal{R}$, $w(x^k) \in \mathcal{R}^+$ and is bilinear in $y(x^k)$ and $w(x^k)$. The quantity

$$P(x^k; y; w) = \partial_m P^m(x^k; y; w) \tag{1.99}$$

is referred to as the *bilinear concomitant* of the operator $\mathcal{O}$. If $\mathcal{O}$ is such that

$$\mathcal{O}y = \mathcal{O}^+y \tag{1.100}$$

for all $y(x^k) \in \mathcal{R} \cap \mathcal{R}^+$, then $\mathcal{O}$ is said to be *self-adjoint*. If

$$\mathcal{O}y = -\mathcal{O}^+y \tag{1.101}$$

for all $y(x^k) \in \mathcal{R} \cap \mathcal{R}^+$, then $\mathcal{O}$ is said to be *anti self-adjoint*. If the boundary conditions, which are satisfied by $y \in \mathcal{R}$ and $w \in \mathcal{R}^+$, are such that $\int_{\partial D} P^m dS_m(x) = 0$ and (1.100) holds, then *the problem is weakly self-adjoint*. If $P^m|_{\partial D} N_m = 0$ and (1.100) holds, where $N_m$ is the field of normal vectors to $\partial D$, then the *problem is self-adjoint*. Any self-adjoint problem is weakly self-adjoint, but not every weakly self-adjoint problem is a self-adjoint problem. It is important to distinguish between self-adjoint operators and self-adjoint problems, for otherwise, seemingly good results will prove useless.

**Theorem 1.7.** *If the Euler-Lagrange operator applied to a lagrangian function* L *of only one dependent function* $y(x^k)$ *gives a linear operator on* $y(x^k)$, *this linear operator is self-adjoint.*

**Proof.** By Theorem 1.6 and the linearity of the Euler-Lagrange operator with respect to lagrangian functions, it is sufficient to establish the results for cases 5, 6, 8 and 9. We shall establish the result for case 9. The other three cases are left to the student as exercises. From (1.75)-(1.77) we have

$$\left\{ \mathcal{E} \mid ak_2{}^2 \right\}_y (x^k) = \int_{D^*} \hat{H}^m(z^k, x^k)\, \partial_m y(z^k)\, dV(z)\,.$$

Hence, with $\mathcal{O}y = \int_{D^*} \hat{H}^m(z^k, x^k)\,\partial_m y(z^k)\,dV(z)$, we obtain

$$\int_{D^*} w\mathcal{O}y\,dV(x) = \int_{D^*}\int_{D^*} w(x^k)\,\hat{H}^m(z^k, x^k)\,\partial_m y(z^k)\,dV(z)\,dV(x).$$

Now,

$$w(x^k)\,\hat{H}^m(z^k, x^k)\,\partial_m y(z^k) = \frac{\partial}{\partial z^m}[w(x^k)\,\hat{H}^m(z^k, x^k)\,y(z^k)] - w(x^k)\,\frac{\partial \hat{H}^m(z^k, x^k)}{\partial z^m}\,y(z^k),$$

When (1.77) is used, we obtain

$$\begin{aligned} w(x^k)\,H^m(z^k, x^k)\,\partial_m y(z^k) &= \frac{\partial}{\partial z^m}[w(x^k)\,\hat{H}^m(z^k, x^k)\,y(z^k)] - w(x^k)\,\frac{\partial \hat{H}^m(x^k, z^k)}{\partial x^m}\,y(z^k) \\ &= \frac{\partial}{\partial z^m}[w(x^k)\,\hat{H}^m(z^k, x^k)\,y(z^k)] - \frac{\partial}{\partial x^m}[w(x^k)\,\hat{H}^m(x^k, z^k)\,y(z^k)] \\ &\quad + \partial_m w(x^k)\,\hat{H}^m(x^k, z^k)\,y(z^k), \end{aligned}$$

and thus,

$$\begin{aligned} &\int_{D^*}\int_{D^*} w(x^k)\,\hat{H}^m(z^k, x^k)\,\partial_m y(z^k)\,dV(z)\,dV(x) \\ &\quad = \int_{D^*}\int_{D^*} \partial_m w(x^k)\,\hat{H}^m(x^k, z^k)\,y(z^k)\,dV(z)\,dV(x) \\ &\quad + \int_{D^*}\int_{D^*} \left\{\frac{\partial}{\partial z^m}[w(x^k)\,\hat{H}^m(z^k, x^k)\,y(z^k)] \right. \\ &\quad \left. - \frac{\partial}{\partial x^m}[w(x^k)\,\hat{H}^m(x^k, z^k)\,y(z^k)]\right\} dV(z)\,dV(x). \end{aligned}$$

An interchange of the integration variables $z$ and $x$ (an allowable operation since their domains of integration are the same) and the divergence theorem gives

$$\int_{D^*} (w\mathcal{O}y - y\mathcal{O}w)\, dV(x)$$

$$= \int_{\partial D^*} \left\{ \int_{D^*} H^m(x^k, z^k)[y(z^k)\, w(x^k) - y(x^k)\, w(z^k)]\, dV(z) \right\} dS_m(x).$$

If a linear operator involves only differentiation, then $P^m(x^k; y; w)$ can be computed directly without integration. Such operators are said to have a *localizable bilinear concomitant*. In the above example, the linear operator involved an integration process, the calculation involved an integration over $D^*$ in order to interchange the $x$ and $z$ integration variables. In such cases, the operator is said to have a *nonlocalizable bilinear concomitant*.

A linear operator is said to be of *second order* if and only if it involves derivatives of order less than three and integrals of derivatives of order less than two.

**Theorem 1.8.** *The most general self-adjoint integrodifferential equation of second order (that is* $\mathcal{O}y + f = 0$, *where* $\mathcal{O}$ *is a second-order linerar operator) is the Euler equation obtained from the lagrangian function given by* (1.96).

**Proof.** The proof is a straightforward application of the definition of a linear operator of second order and the results established above. Alternatively, one can simply write the most general linear operator of second order and demand that it be self-adjoint. The result of this requirement is the Euler equation for the lagrangian equation (1.96).

## 1.8. THE VARIATIONAL EMBEDDING PROBLEM

Suppose that we are given an integrodifferential equation for the determination of a function $v(x^k)$ on a domain $D^*$ of $X_n$. The variational embedding problem involves finding when this given integrodifferential equation is the Euler equation for a lagrangian function. Specifically, the problem involves finding a functional on $\mathfrak{D}_1(D^*)$ such that the stationarization set of this functional is comprised of solutions of the given integrodifferential equation.

If we use the results of Sec. 1.7, the following theorem is immediate.

**Theorem 1.9.** *Any linear integrodifferential equation of the form*

$$\mathcal{O}y + f = 0,$$

*where* $\mathcal{O}$ *is a self-adjoint linear operator of second order, can be embedded in a variational statement with one dependent function* $y(x^k)$.

By considering lagrangian functions which depend on derivatives higher than the first, it can be shown that any self-adjoint linear operator equation can be embedded in a variational statement. We have restricted out consideration to problems for which the lagrangian function depends only on first derivatives since most equations of interest are of order less than three.

Consider the simple situation in which the given equation is

$$-\frac{\partial y}{\partial x^1} + \frac{\partial^2 y}{\partial (x^2)^2} + \frac{\partial^2 y}{\partial (x^3)^2} = 0.$$

This equation is not expressable in terms of a self-adjoint operator and, hence, cannot be embedded in a variational statement of the kind considered above. Since it is linear and involves derivatives of order less than three, some lagrangian function involving only first derivatives should lead to this equation. This is the case, but the lagrangian function must depend on at least two unknown functions, not just one. For this reason, and for the equally valid reason that systems of equations are of considerable interest, we take up the study of lagrangian functions of several dependent functions in Chap. 2.

There are a number of reasons for wanting to embed a given integrodifferential equation in a variational statement. First, from the physical point of view, the variational embedding always gives a lagrangian function, and a problem in which the integrodifferential equation arises can be viewed in terms of an energy structure. This follows from the fact that the lagrangian function of a physical system can usually be written as the difference between the kinetic energy density of the system and the potential energy density of the system—the functional $J[y](L)$ is interpreted as the total action of the system. The ability to view the problem in these terms often leads to significant insight into the original integrodifferential equation and the properties of its solutions. Second, a variational statement can often be used to establish the existence of solutions to the given integrodifferential equation; and third, a variational statement usually leads to a very efficient method of constructing approximate solutions.

Suppose that we could show that any stationarizing element of $\mathfrak{D}_1(D^*)$ also minimizes the value of $J[y](L)$, that there are elements of $\mathfrak{D}_1(D^*)$ for which $J[y](L) < \infty$, and that

$$\inf_{y \in \mathfrak{D}_1(D^*)} J[y](L) = m > -\infty.$$

Then, by the definition of $m$, there exists an infinite sequence $\{y_n(x^k)\}$, called a *minimizing sequence*, such that $y_n(x^k)$ belongs to $\mathfrak{D}_1(D^*)$ for every value of the index $n$ and

$$\lim_{n\to\infty} J[y_n](L) = m.$$

If the sequence $\{y_n(x^k)\}$ has a limit $y_0(x^k)$ that belongs to $\mathfrak{D}_1(D^*)$ and if it is legitimate to write

$$J[y_0](L) = \lim_{n\to\infty} J[y_n](L),$$

that is,

$$J\left[\lim_{n\to\infty} y_n\right](L) = \lim_{n\to\infty} J[y_n](L), \tag{1.102}$$

then we have

$$J[y_0](L) = m.$$

Thus, $y_0(x^m)$ is a solution of the variation problem, and the functions of the minimizing sequence $\{y_n(x^k)\}$ can be regarded as approximate solutions of the variational problem. We may relax the condition that $y(x^k)$ belongs to $\mathfrak{D}_1(D^*)$ and replace it by the condition that $y(x^k)$ belongs to some normed linear space of functions. The norm of the linear space must be such, however, that we can establish the validity of (1.102). In this context, we state the following result, the proof of which is an elementary exercise in $\epsilon$ manipulations.

**Lemma 1.10.** *If $\{y_n(x^k)\}$ is a minimizing sequence of the functional $J[y](L)$ with limit function $y_0(x^k)$, and if $J[y](L)$ is lower semicontinuous* *at* $y_0(x^k)$, *then*

$$J[y_0](L) = \lim_{n \to \infty} J[y_n](L).$$

The convergence and lower semicontinuity are, of course, to be calculated with respect to the norm of the function space from which $y_n$ comes. It should also be noted that $y_0(x^k)$ must belong to the function space since we require establishment of lower semicontinuity at $y_0(x^k)$.

The following observation is trivial yet very important and, thus, is specifically stated.

*The variational embedding of an integrodifferential equation is never unique.* Specifically, *if* L *is a lagrangian function that embeds the given integrodifferential equation, then any lagrangian function in the equivalence class* $\hat{L}$ *that contains* L *also embeds the given integrodifferential equation in a variational statement.* An interesting, and as yet unresolved question, is whether the rapidity of convergence of a minimizing sequence changes when $J[y]('L)$ is used instead of $J[y](L)$, in which $'L = L + A$ for $A \in \mathfrak{N}$. As a trivial example, if the inf over $y(x^k) \in \mathfrak{D}_1(D^*)$ of $J[y](L)$ is the finite number $m$, then the inf over $y(x^k) \in \mathfrak{D}_1(D^*)$ of $J[y](L + u)$ is zero if we take $u(x^k)$ to be such that $\int_{D^*} u(x^k)\, dV(x) = -m$. A nontrivial question is involved here, for the use of elements of $\mathfrak{N}$ that do not belong to $\mathfrak{N}^*$ can change the values of the functional under consideration in any continuous manner. If we are concerned with a linear problem, the lagrangian function has to be quadratic, and $A \in \mathfrak{N}$ must be quadratic at most in $y(x^k)$. If this structure is used, we can obtain any system of nonnatural boundary conditions which follow the pattern given in Sec. 1.9 without considering added surface functionals.

## 1.9. BOUNDARY CONDITIONS AND CONTINUITY REQUIREMENTS

If $y(x^k) \in \mathcal{C}(D^*)$, then the values of $y(x^k)$ are well defined when $\{x^k\}$ is restricted to $\partial D$ (recall that $D^*$ is the closure of $D$ in $X_n$ and is assumed to be compact). If $y(x^k) \in \mathfrak{D}_1(D^*)$, then the values of $\partial_m y(x^k)$ are well defined on $\partial D$ as being the limits of $\partial_m y(x^k)$ as $\partial D$ is approached from the interior of $D^*$.

Let $s_b^{\ m}\left[x^k, z^k, y(z^k), \partial_t y(z^k)\right]$, $b = 1, \ldots, p$ be $np$ functions defined for all $\{x^k\} \in \partial D$ and all $z^k \in \partial D$. Let it also be continuous for all $\{x^k\}$ and $\{z^k\}$ on $\partial D$ and be of class $C^1$ in $y(z^k)$ and $\partial_t y(z^k)$. With these functions we can define $p$ functionals by the relations

$$s_b(x^m; y) = \int_{\partial D} s_b{}^m\left[x^k, z^k, y(z^k), \partial_t y(z^k)\right] dS_m(z) \qquad (1.103)$$

for all $\{x^k\} \in \partial D$.

Let $L^m\left[x^k, y(x^k), \partial_t y(x^k), s_b(x^k; y)\right]$ be $n$ functions defined for all $\{x^k\} \in \partial D$ and be of class $C^1$ in $y(x^k)$, $\partial_t y(x^k)$ and $s_b(x^k; y)$. If $J[y](L)$ is as previously defined, we consider the composite functional $J[y](L, L^m)$ defined by

$$J[y](L, L^m) = J[y](L) + \int_{\partial D} L^m\left[x^k, y(x^k), \partial_t y(x^k), s_b(x^k; y)\right] dS_m(x). \qquad (1.104)$$

*If* $L^k$ *and* $s_b{}^k$ *are each independent of* $\partial_t y(x^k)$, $\partial_t y(z^k)$, *then an element* $y(x^k)$ *of* $\mathfrak{D}_1(D^*)$ *is said to stationarize* $J[y](L, L^m)$ *if and only if*

$$\delta J[y](L, L^m) = 0 \qquad (1.105)$$

*for all variations* $\delta y(x^k) \in \mathfrak{D}_1(D^*)$, not just for those variations which vanish on the boundary of $D^*$. The above conditions on $L^k$ and $s_b{}^k$ most often occur in practice. For the general case, an element $y(x^k)$ of $\mathfrak{D}_1(D^*)$ is said to *stationarize* $J[y](L, L^m)$ *if and only if* (1.105) *is satisfied for all variations* $\delta y(x^k) \in \mathfrak{D}_1(D^*)$ *such that* $\partial_m \delta y(x^k)\big|_{\partial D} = 0$. In either case, the collection of all stationarizing elements is denoted by $S(L, L^m)$.

We now must obtain conditions such that $\delta J[y](L, L^m) = 0$. From the definition of $\Delta J$ and $J[y](L, L^m)$, we have

$$\begin{aligned}\Delta J[y](L, L^m) = {} & \Delta J[y](L) \\ & + \int_{\partial D} L^m\left[x^k, y + \delta y, \partial_t y + \partial_t \delta y, s_b(x^k; y + \delta y)\right] dS_m(x) \\ & - \int_{\partial D} L^m\left[x^k, y, \partial_t y, s_b(x^k; y)\right] dS_m.\end{aligned} \qquad (1.106)$$

A modification of the results of Sec. 1.5, which account for the fact that $\delta y(x^k)$ is no longer required to vanish on $\partial D$, gives

$$k_a(x^k; y + \delta y) = k_a(x^k; y) + \delta k_a(x^k; y) + o\left(\|\delta y\|\right), \qquad (1.107)$$

where

$$k_a(x^k;\, y) = \int_{D^*}\left\{\frac{\partial g_a}{\partial y(z)} - \frac{\partial}{\partial z^m}\left[\frac{\partial g_a}{\partial(\partial_m y(z))}\right]\right\}\delta y(z)\, dV(z)$$
$$+ \int_{D^*}\frac{\partial}{\partial z^m}\left[\frac{\partial g_a}{\partial(\partial_m y(z))}\,\delta y(z)\right] dV(z)\,. \tag{1.108}$$

The modification also gives that $y(z^k)$ must be of class $C^2$ if

$$\partial^2 g_a/\partial(\partial_n y)\,\partial(\partial_m y) \neq 0.$$

It is thus evident that

$$\Delta J[y](L) = \int_{D^*}\left\{\frac{\partial L}{\partial y}\delta y + \frac{\partial L}{\partial(\partial_m y)}\partial_m \delta y + \frac{\partial L}{\partial k_a}\delta k_a\right\} dV(x) + o\big(\|\delta y\|\big)$$
$$= \int_{D^*}\left\{\frac{\partial L}{\partial y}\delta y - \frac{\partial}{\partial x^m}\left[\frac{\partial L}{\partial(\partial_m y)}\right]\delta y + \frac{\partial L}{\partial k_a}\delta y\right\} dV(x) \tag{1.109}$$
$$+ \int_{D^*}\frac{\partial}{\partial x^m}\left[\frac{\partial L}{\partial(\partial_m y)}\delta y\right] dV(x) + o\big(\|y\|\big)$$

and that $y(x^k)$ must be of class $C^2$ unless $\partial^2 L/\partial(\partial_m y)\,\partial(\partial_n y) = 0$. When (1.108) is used, we have

$$\int_{D^*}\frac{\partial L}{\partial k_a}\delta k_a\, dV(x) = \int_{D^*}\int_{D^*}\frac{\partial L}{\partial k_a}(x)\left\{e\,|\,g_a(x^k)\right\}_y(z^k)\,\delta y(z)\, dV(z)\, dV(x)$$
$$+ \int_{D^*}\int_{D^*}\frac{\partial L}{\partial k_a}(x)\frac{\partial}{\partial z^m}\left[\frac{\partial g_a}{\partial(\partial_m y(z))}\delta y(z)\right] dV(z)\, dV(x) + o\big(\|\delta y\|\big)\,.$$

Consequently, we have

$$\delta J[y](L) = \int_{D^*}\{\mathcal{E}|L\}_y(x^k)\,\delta y(x)\, dV(x) + \int_{D^*}\left\{\frac{\partial}{\partial x^m}\left[\frac{\partial L}{\partial(\partial_m y)}\delta y(x)\right]\right.$$
$$\left. + \int_{D^*}\frac{\partial L}{\partial k_a}(x^k)\,\frac{\partial}{\partial z^m}\left[\frac{\partial g_a}{\partial(\partial_m y(z))}\delta y(z)\right] dV(z)\right\} dV(x)\,. \tag{1.110}$$

Thus, an interchange of $x^k$ and $z^k$ in the last integral gives

$$\delta J[y](L) = \int_{D^*} \{\mathcal{E} \mid L\}_y (x^k)\, \delta y(x)\, dV(x)$$

$$+ \int_{D^*} \frac{\partial}{\partial x^m} \left\{ \left[ \frac{\partial L}{\partial(\partial_m y)} + \int_{D^*} \frac{\partial L}{\partial k_a}(z) \frac{\partial g_a^*}{\partial(\partial_m y)}\, dV(z) \right] \delta y(x) \right\} dV(x). \tag{1.111}$$

A straightforward operation based on (1.103) yields

$$\delta s_b(x^k; y) = \int_{\partial D^*} \left[ \frac{\partial s_b{}^m}{\partial y(z)} \delta y(z) + \frac{\partial s_b{}^m}{\partial(\partial_t y(z))} \partial_t \delta y(z) \right] dS_m(z). \tag{1.112}$$

Thus,

$$\Delta \int_{\partial D} L^m dS_m(x) = \int_{\partial D} \left[ \frac{\partial L^m}{\partial y} \delta y + \frac{\partial L^m}{\partial(\partial_t y)} \partial_t \delta y + \frac{\partial L^m}{\partial s_b} \delta s_b \right] dS_m + o(\|\delta y\|)$$

$$= \int_{\partial D} \left[ \frac{\partial L^m}{\partial y} \delta y + \frac{\partial L^m}{\partial(\partial_t y)} \partial_t \delta y \right] dS_m(x) + o(\|\delta y\|)$$

$$+ \int_{\partial D} \int_{\partial D} \frac{\partial L^m}{\partial s_b}(x) \left[ \frac{\partial s_b{}^t}{\partial y(z)} \delta y(z) + \frac{\partial s_b{}^t}{\partial(\partial_r y(z))} \partial_r \delta y(z) \right] dS_t(z)\, dS_m(x).$$

Because variations of surface integrals are now under consideration, $\mathfrak{D}_1(D^*)$ must be modified to include the additional terms

$$\max_{x \in \partial D} |y(x^k)| \qquad \sum_{r=1}^{n} \max_{x \in \partial D} |\partial_r y(x^k)|.$$

This was assumed in the above expressions by writing $o(\|\delta y\|)$. Finally, reversing $x^k$ and $z^k$ in the last double surface integral and using the notation introduced in Sec. 1.5, we have

$$\delta \int_{\partial D} L^m dS_m(x) = \int_{\partial D} \left[ \frac{\partial L^m}{\partial y} + \int_{\partial D} \frac{\partial L^t}{\partial s_b}(z) \frac{\partial s_b^{*m}}{\partial y}\, dS_t(z) \right] \delta y(x)\, dS_m(x)$$

$$+ \int_{\partial D} \left[ \frac{\partial L^m}{\partial(\partial_r y)} + \int_{\partial D} \frac{\partial L^t}{\partial s_b}(z) \frac{\partial s_b^{*m}}{\partial(\partial_r y)}\, dS_t(z) \right] \partial_r \delta y(x)\, dS_m(x). \tag{1.113}$$

If $L^m$ and $s_b{}^m$ do not depend on the derivatives of $y$, the last integral over the boundary of $D^*$ vanishes identically. If this is not the case, we require that $\partial_m \delta y(x^k)\big|_{\partial D} = 0$. In either case, we have

$$\delta \int_{\partial D} L^m dS_m(x) = \int_{\partial D} \left[ \frac{\partial L^m}{\partial y} + \int_{\partial D} \frac{\partial L^t}{\partial s_b}(z) \frac{\partial s_b^{*m}}{\partial y} dS_t(z) \right] \delta y(x)\, dS_m(x) . \quad (1.114)$$

Under the assumed continuity conditions, the divergence theorem can be applied to the second integral over $D^*$ in (1.111) giving

$$\begin{aligned} \delta J[y](L, L^m) = & \int_{D^*} \{\mathcal{E} \mid L\}_y (x^k)\, \delta y(x^k)\, dV(x) \\ & + \int_{\partial D} \left[ \frac{\partial L}{\partial(\partial_m y)} + \frac{\partial L^m}{\partial y} + \int_{D^*} \frac{\partial L}{\partial k_a}(z) \frac{\partial g_a^*}{\partial(\partial_m y)} dV(z) \right. \\ & \left. + \int_{\partial D} \frac{\partial L^t}{\partial s_b}(z) \frac{\partial s_b^{*m}}{\partial y} dS_t(z) \right] \delta y(x)\, dS_m(x) . \end{aligned} \quad (1.115)$$

A subclass of all variations, or of all variations whose derivatives vanish on the boundary of $D^*$, consists of those variations which vanish on $\partial D^*$. Hence, by Lemma 1.2, $\delta J[y](L, L^m) = 0$, only if $\{\mathcal{E} \mid L\}_y (x^k) = 0$ at all interior points of $D^*$. Also $y(x^k)$ must be of class $C^2$ except at those points at which

$$\frac{\partial^2 L}{\partial(\partial_n y)\, \partial(\partial_m y)} + \int_{D^*} \frac{\partial L}{\partial k_a}(z) \frac{\partial^2 g_a^*}{\partial(\partial_n y)\, \partial(\partial_m y)} dV(z) = 0 .$$

Satisfaction of this condition now gives

$$\begin{aligned} \delta J[y](L, L^m) = & \int_{\partial D} \left[ \frac{\partial L}{\partial(\partial_m y)} + \frac{\partial L^m}{\partial y} + \int_{D^*} \frac{\partial L}{\partial k_a}(z) \frac{\partial g_a^*}{\partial(\partial_m y)} dV(z) \right. \\ & \left. + \int_{\partial D} \frac{\partial L^t}{\partial s_b}(z) \frac{\partial s_b^{*m}}{\partial y} dS_t(z) \right] \delta y(x^k)\, dS_m(x) . \end{aligned} \quad (1.116)$$

we write

$$dS_m(x) = N_m(x)\, dS(x) , \quad (1.117)$$

where $\{N_m(x)\}$ is the unit normal vector field of $\partial D$ with the orientation from $D$ to the complement of $D^*$ and $dS(x)$ is the differential element of surface measure on $\partial D$. Application of Lemma 1.1 to the case in which the integration is over the $n$-1 dimensional region $\partial D$, gives the condition

$$N_m(x)\left[\frac{\partial L}{\partial(\partial_m y)} + \frac{\partial L^m}{\partial y} + \int_{D^*} \frac{\partial L}{\partial k_a}(z)\,\frac{\partial g_a^*}{\partial(\partial_m y)}\,dV(z) + \int_{\partial D} \frac{\partial L^t}{\partial s_b}(z)\,\frac{\partial s_b^{*m}}{\partial y}\,dS_t(z)\right]\Bigg|_{\partial D} = 0 \tag{1.118}$$

or $\delta y(x^m)\big|_{\partial D} = 0$.

**Theorem 1.10.** *If* $y(x^k)$ *is of class* $C^2$ *at all interior points of* $D^*$, *except those at which*

$$\frac{\partial^2 L}{\partial(\partial_n y)\,\partial(\partial_m y)} + \int_{D^*} \frac{\partial L}{\partial k_a}(z)\,\frac{\partial^2 g_a^*}{\partial(\partial_n y)\,\partial(\partial_m y)}\,dV(z) = 0\,,$$

*then* $y(x^k)$ *belongs to* $S(L, L^m)$ *if and only if* $y(x^k)$ *satisfies the Euler equation at all interior points of* $D^*$ *and either* (1.118) *holds on* $\partial D$ *or* $\delta y(x^k)\big|_{\partial D} = 0$ *on* $D$. *The latter condition can be weakened to read either* (1.118) or $\delta y(x^k) = 0$ *holds at each point of* $\partial D$.

When $L^m \equiv 0$, (1.118) reduces to

$$N_m(x)\left[\frac{\partial L}{\partial(\partial_m y)} + \int_{D^*} \frac{\partial L}{\partial k_a}(z)\,\frac{\partial g_a^*}{\partial(\partial_m y)}\,dV(z)\right]\Bigg|_{\partial D} = 0\,. \tag{1.119}$$

These conditions are referred to as the *natural boundary conditions* associated with the variational problem. If they are not satisfied, the additional functional on $\partial D$ must be included in order to obtain (1.118). The added terms in (1.118) which do not occur in (1.119) may thus be viewed as compensating terms arising from the non-natural boundary conditions of the problem. The student should verify that any problem which satisfies natural boundary conditions and comes from a lagrangian function of the form of Cases 1, 5, 6, 8 and 9 of Sec. 1.7 will be a self-adjoint problem.

It is evident, that the continuity assumptions on the functions $L$, $L^m$, $g_a$ and $s_b{}^m$ can be relaxed. The extent to which they can be relaxed in any particular case is left to the student. An important point arises in this context. We have repeatedly used the divergence theorem to convert the volume integral of a divergence into a surface integral, assuming the continuous differentiability of all quantities that make up the divergence. If $L$ or $g_a^*$ is not of class $C^2$ in $\partial_m y(x^k)$ then some of the quantities forming the divergence are not differentiable throughout $D^*$. In this event, we must use the modified divergence theorem.

Let $D$ be divided into two parts $D^+$ and $D^-$ by an $(n-1)$-dimensional surface $\pi$, so that $D^* = (D^+ \cup D^-)^*$. Let $f^m(x^k)$ be a function of class $C^1$ on $D^+$ and on $D^-$ and with limiting values on $\pi$. The divergence theorem then reads

$$\int_{D^*} \partial_m f^m(x^k)\, dV(x) + \int_{D^* \cap \pi} [\![ f^m ]\!] J_m\, d\pi = \int_{D^* - (D^+ \cup D^- \cup \pi)} f^m(x^k)\, dS_m(x)$$

where

$$[\![ f^m ]\!] = f^{m+} - f^{m-} \quad , \quad f^{m+} = \lim_{\substack{x \to \pi \\ x \in D^+}} f^m \quad , \quad f^{m-} = \lim_{\substack{x \to \pi \\ x \in D^-}} f^m \quad ,$$

and $J_m d\pi$ is the directed differential element of surface with the orientation from $D^-$ to $D^+$. The student is advised to obtain the conditions that must hold for $\delta J[y](L) = 0$ when $\partial L / \partial(\partial_m y)$ is piecewise continuous on $D^*$.

If we are concerned with a linear problem, we must then restrict the possible form of $L^m$ to the following (functionals ignored for simplicity):

$$L^m = u^m(x^k) + v^m(x^k)\, y(x^k) + \frac{1}{2} w^m(x^k)\, y(x^k)^2 \, . \tag{1.120}$$

We also know that we can add any element of $\mathfrak{N}$ to $L$ without changing the Euler equation. If the problem is to remain linear, we can add

$$\begin{aligned} A = {} & \partial_m^* a^m(x^k) + y(x^k)\, \partial_m^* b^m(x^k) + b^m(x^k)\, \partial_m y(x^k) \\ & + \frac{1}{2} y(x^k)^2\, \partial_m^* c^m(x^k) + c^m(x^k)\, y(x^k)\, \partial_m y(x^k) \, , \end{aligned} \tag{1.121}$$

a divergence. Hence, if we substitute $L + A$ for $L$ into (1.118) and use (1.120) and (1.121), we obtain

$$N_m(x) \left[ \frac{\partial L}{\partial(\partial_m y)} + b^m + c^m y + v^m + w^m y \right] \Bigg|_{\partial D} = 0 \, . \tag{1.122}$$

It thus follows that we can always achieve the desired purpose in the linear case by adding the appropriate element of $\mathfrak{N}$ to $L$ rather than adding an integral over $\partial D$ to $J[y](L)$. In one case we have $b^m = c^m = 0$ with given $v^m$ and $w^m$. In the other we first set $b^m = v^m$, $c^m = w^m$ and then discard the surface integral which had been added to $J[y](L)$. It follows that use of *given boundary conditions can provide a means of selecting a unique element of* $\hat{L}$ *for a given Euler equation.*

We now come to a very important application of the nonlocal variational calculus and stationarization. Set $a = 1$ and

$$k_1(x^m; y) = k(y) = \int_{D*} y(z)^2 \, dV(z) , \tag{1.123}$$

so that

$$g_1 = g = y(z)^2 , \qquad g_1^* = g^* = y(x)^2 . \tag{1.124}$$

The lagrangian function is taken as

$$L = k^{-1} \delta^{mn} \partial_m y \, \partial_n y = k^{-1} \vec{\nabla} y \cdot \vec{\nabla} y , \tag{1.125}$$

and, hence, the functional

$$J[y](L) = \int_{D*} L \, dV(x) = \frac{\int_{D*} \vec{\nabla} y \cdot \vec{\nabla} y \, dV(x)}{\int_{D*} y(z)^2 \, dV(z)} \tag{1.126}$$

is of the form of a Rayleigh quotient, the stationarization of which can be handled only within the classical (local) variational calculus with great difficulty. With the nonlocal variational calculus the matter is quite simple. We have

$$\frac{\partial L}{\partial y(x)} = 0 , \qquad \frac{\partial L}{\partial(\partial_m y(x))} = 2k^{-1} \partial_m y(x) ,$$

$$\frac{\partial L}{\partial k}(x) = -k^{-2} \vec{\nabla} y(x) \cdot \vec{\nabla} y(x) , \qquad \frac{\partial L}{\partial k}(z) = -k^{-2} \vec{\nabla} y(z) \cdot \vec{\nabla} y(z) , \tag{1.127}$$

$$\frac{\partial g^*}{\partial y(x)} = 2y(x) , \qquad \frac{\partial g^*}{\partial(\partial_m y(x))} = 0 ,$$

and hence

$$\{\mathcal{E}\,|\,L\}_y(x) = -2k^{-1}\left[\nabla^2 y(x) + y(x)\,k^{-1}\int_{D^*} \vec{\nabla} y(z)\cdot\vec{\nabla} y(z)\,dV(z)\right]$$

where

$$\nabla^2 y(x) = \sum_{m=1}^{n} \frac{\partial^2 y(x)}{\partial x^m \partial x^m}.$$

If we exclude the possibility $y(x) \equiv 0$, so that $k(y) > 0$, and use (1.126), then $y(x)$ stationarizes $J[y]$ if and only if

$$\nabla^2 y(x) = -J[y]\,y(x). \tag{1.128}$$

The function $y(x)$ that stationarizes the Rayleigh quotient (1.126) is an eigenfunction of Laplace's operator (a solution of Helmholtz' equation) with the associated eigenvalue given by $-J[y]$. The natural boundary conditions associated with (1.126) are

$$y(x)\big|_{\partial D} = f(x)\big|_{\partial D} \tag{1.129}$$

or

$$\left.\frac{\partial L}{\partial(\partial_m y(x))}\right|_{\partial D} N^m \equiv 2k^{-1}\,\partial_m y(x)\big|_{\partial D} N^m = 0 \tag{1.130}$$

If, instead of (1.126), we take

$$J_1[y] = J[y] + \int_{D^*} \partial_m V^m[x, y(x)]\,dV(x), \tag{1.131}$$

then $y(x)$ still satisfies (1.128) but the boundary condition (1.130) is replaced by

$$2\partial_m y(x)\big|_{\partial D} N^m = -\left.\frac{\partial V^m(x,y)}{\partial y}\right|_{\partial D} N^m k. \tag{1.132}$$

Thus the functional representation of the eigenvalues of Laplace's operator, given by $-J[y]$, is the same for the natural boundary conditions (1.130) and for the nonnatural boundary conditions (1.132). Particular note should be taken of the fact that the boundary condition (1.132) is nonlocal:

$$2\partial_m y(x)\Big|_{\partial D} N^m = \frac{\partial V^m(x,y)}{\partial y}\Big|_{\partial D} N^m \int_{D^*} y(z)^2\, dV(z)\,.$$

To obtain local boundary conditions, we must take

$$J_2[y] = J[y] + k^{-1} \int_{D^*} \partial_m V^m[x, y(x)]\, dV(x)\,, \tag{1.133}$$

in which case the Euler equation is

$$\nabla^2 y(x) = -J_2[y]\, y(x) \tag{1.134}$$

and the boundary conditions are

$$2\partial_m y(x)\Big|_{\partial D} N^m = -\frac{\partial V^m(x,y)}{\partial y}\Big|_{\partial D} N^m\,. \tag{1.135}$$

Theorem 1.11 now becomes evident by straightforward generalization of the above results.

**Theorem 1.11.** *If* $y(x)$ *is such that it stationarizes the generalized Rayleigh quotient*

$$J[y] = k^{-1} \int_{D^*} F\left[x^m, y(x^m), \partial_t y(x^m)\right] dV(x)\,, \tag{1.136}$$

*where* $F$ *is a function of class* $C^2$ *in its indicated arguments and*

$$k = \int_{D^*} y(z)^2\, dV(z)\,, \tag{1.137}$$

*then* $y(x)$ *is an eigenfunction of*

$$\{e \mid F\}_y(x) \tag{1.138}$$

*with the associated eigenvalue* $2J[y]$ *(that is,*

$$\{e \mid F\}_y(x) = 2J[y]\, y(x) \quad ). \tag{1.139}$$

*The natural boundary conditions are*

$$y(x)\big|_{\partial D} = f(x)\big|_{\partial D} \tag{1.140}$$

*or*

$$k^{-1} \frac{\partial F}{\partial(\partial_m y)}\bigg|_{\partial D} N^m = 0 . \tag{1.141}$$

*If the boundary conditions are nonnatural, then either*

$$J_1[y] = J[y] + \int_{D^*} \partial_m V^m[x^t, y(x^t)]\, dV(x) \tag{1.142}$$

*is stationary so that* $y(x)$ *satisfies* (1.139) *and the boundary conditions are nonlocal,*

$$k^{-1} \frac{\partial F}{\partial(\partial_m y)}\bigg|_{\partial D} N^m = -\frac{\partial V^m(x, y)}{\partial y}\bigg|_{\partial D} N^m , \tag{1.143}$$

*or*

$$J_2[y] = J[y] + k^{-1} \int_{D^*} \partial_m V^m[x^t, y(x^t)]\, dV(x) \tag{1.144}$$

*is stationary so that* $y(x)$ *satisfies*

$$\{e \,|\, F\}_y(x) = 2J_2[y]\, y(x) \tag{1.145}$$

*and the boundary conditions are local,*

$$\frac{\partial F}{\partial(\partial_m y)}\bigg|_{\partial D} N^m = \frac{\partial V^m(x, y)}{\partial y}\bigg|_{\partial D} N^m . \tag{1.146}$$

Theorem 1.11 makes no assumption concerning whether $\{e \,|\, F\}_y(x)$ is or is not a linear operator (on $y(x)$). Thus, nonlinear as well as linear operators are included in this theorem. Further, if $\{e \,|\, F\}_y(x)$ is either a linear or a homogeneous operator, $\{e \,|\, F\}_y(x)$ is formally self-adjoint. In view of the results concerning the variational embedding of linear self-adjoint operators of second order and since the function $F$ that accomplishes the embedding is obtained easily, Theorem 1.11 provides a variational formulation for the eigenvalues and eigenfunctions of such operators.

*PROBLEMS*

1.1. Construct a function of one variable on a closed interval $D^*$ which is of class $C^1$ and such that the $\mathcal{C}(D^*)$ norm of the function is less than $\epsilon$ while the $\mathfrak{D}_1(D^*)$ norm of the function is greater than 10.

1.2. Let $F(x^m, p, q_m)$ be a continuous function of its $2n+1$ arguments and define the functional $J[y](F)$ by the relation

$$J[y](F) = \int_{D^*} F\left[x^k, y(x^k), \partial_m y(x^k)\right] dV(x),$$

in which $D^*$ is a simply connected, compact point set in $n$-dimensional space and $y(x^k)$ is an element of $\mathfrak{D}_1(D^*)$. Show that $J[y](F)$ is continuous with respect to the $\mathfrak{D}_1(D^*)$ norm and that it is not continuous with respect to the $\mathcal{C}(D^*)$ norm.

1.3. Let $\mathfrak{D}_2(D^*)$ denote the space of functions of class $C^2$ on $D^*$ with the uniform convergence norm with respect to $y(x^k)$ and each of its first and second derivatives. If $J[y](F)$ is defined by

$$(*) \quad J[y](F) = \int_{D^*} F\left[x^k, y(x^k), \partial_m(x^k), \partial_n\partial_m y(x^k)\right] dV(x),$$

then show that $J[y](F)$ is differentiable if the function $F$ is of class $C^2$ in its arguments.

1.4. Derive the Euler equation for the functional (*) given in Prob. 1.3.

1.5. Prove that the Euler-Lagrange operator is a linear operator on the class of all lagrangian functions of class $C^2$ in their arguments.

1.6. Give two explicit examples of elements of the nontrivial null class of the Euler-Lagrange operator. These examples are to involve derivatives and integrals of derivatives. Make explicit calculation of the Euler equation for each and show that it is identically satisfied by every element of $\mathfrak{D}_1(D^*)$.

1.7. Work out the Euler equations for all 18 cases given in Sec. 1.7.

1.8. Under what conditions does the integral over $D^*$, involved in the Euler equation of Case 9 in Sec. 1.7, reduce to a surface integral?

1.9. When the continuity conditions on the functions $g_a$ are relaxed so that $\partial g_a/\partial(\partial_m y)$ is discontinuous across some $(n-1)$-dimensional hypersurface in $D^*$, what are the resulting Euler-Lagrange operator and subsidiary conditions? Compute the results in this case for the lagrangian function given in Case 9 when the functions $H^k(z^m, x^m)$ have a discontinuous derivative with respect to at least one $x$ across a hypersurface in $D^*$.

1.10. Prove that the Euler-Lagrange operator applied to the lagrangian functions of Cases 5, 6 and 8 results in self-adjoint operators on $y(x^k)$. Find the bilinear concomitant in each case?

1.11. Let $u$ be a nonzero constant. Obtain the lagrangian whose Euler equation is the scalar wave equation $\frac{1}{u^2}\frac{\partial^2\phi}{\partial t^2} - \frac{\partial^2\phi}{\partial x^2} - \frac{\partial^2\phi}{\partial y^2} - \frac{\partial^2\phi}{\partial z^2} = 0$. Is this lagrangian unique? If $u(t)$ is a function of class $C^1$ in the variable $t$ such that $u(t) > 0$ for all $t$, then find the lagrangian function that yields the generalized wave equation

$$\frac{1}{u^2}\left(\frac{\partial^2\phi}{\partial t^2} - \frac{2}{u}\frac{du}{dt}\frac{\partial\phi}{\partial t}\right) - \frac{\partial^2\phi}{\partial x^2} - \frac{\partial^2\phi}{\partial y^2} - \frac{\partial^2\phi}{\partial z^2} = 0.$$

1.12. Let $c$ and $m$ be nonzero constants. Obtain the lagrangian function whose Euler equation is the Klein-Gordon equation

$$\frac{\partial^2\phi}{\partial x^2} + \frac{\partial^2\phi}{\partial y^2} + \frac{\partial^2\phi}{\partial z^2} - \frac{1}{c^2}\frac{\partial^2\phi}{\partial t^2} - \frac{1}{m^2}\phi = 0.$$

$c$ = speed-of-light and $m = h/2\pi c m_0$, $h$ = Planck's constant and $m_0$ = rest mass.

1.13. Let $f(x^m)$ be a function defined on $D^*$ and nonnegative there. Obtain the lagrangian function whose Euler equation is the generalized Poisson equation $\sum_{i=1}^{n}\left(\partial^2\phi/\partial x^{i\,2}\right) = f$. Obtain the lagrangian function whose Euler equation is the generalized Helmholtz's equation $\sum_{i=1}^{n}\left(\partial^2\phi/\partial x^{i\,2}\right) = \lambda\phi$. Compute the bilinear concomitant for the linear operator in Poisson's and in Helmholtz's equations.

1.14. Find the properties which $K(x^k, z^k)$ must have in order that there exist a lagrangian function of the type considered in this chapter whose Euler equation is the Fredholm integral equation

$$u(x^k) = f(x^k) + \int_{D^*} K(x^k, z^k)\, u(z^k)\, dV(z).$$

What is the lagrangian function under these conditions?

1.15. Find the properties which $K(x^k, z^k)$ and $h^m(x^k, z^k)$ must have in order that there exist a lagrangian function whose Euler equation is the integral equation

$$u(x^k) = f(x^k) + \int_{D^*} K(x^k, z^k)\, u(z^k)\, dV(z) + \int_{D^*} h^m(x^k, z^k)\, \partial_m u(z^k)\, dV(z).$$

What is the lagrangian function under these conditions?

1.16. Work out the most general functions $L^m$ and $s_b{}^m$ for which the surface integrals in $J[y](L, L^m)$ are such that the Euler equation and the boundary conditions are linear. This is the natural completion of the work of Sec. 1.7. In all cases, relate the bilinear concomitant of the linear operator occurring in the Euler equation and the resulting surface term. If $L^m$ is identically zero, show that satisfaction of the boundary conditions for variations which do not vanish on the boundary will always annihilate the bilinear concomitant. Conclude that a self-ajoint operator arising from the variational calculus always gives a self-adjoint problem if the natural (variational) boundary conditions are used.

1.17. Let $L = L\left[x^k, y(x^k), \partial_m y(x^k), \ldots, \partial_{m_1}\partial_{m_2} \cdots \partial_{m_s} y(x^k)\right]$ be of class $C^{2s}$ in its arguments in which $y(x^k) \in C^s$. Find the function space for $J[y](L) = \int_{D^*} L\, dV(x)$, for which $J[y](L)$ is continuous. Show that the Euler-Lagrange operator is given by

$$\{\mathcal{E}|L\}_y(x^k) = \frac{\partial L}{\partial y} + \sum_{i=1}^{s} (-1)^i \frac{\partial^i}{\partial x^{m_1} \partial x^{m_2} \ldots \partial x^{m_i}} \left[\frac{\partial L}{\partial\left(\partial_{m_1} y\right) \partial\left(\partial_{m_2} y\right) \cdots \partial\left(\partial_{m_i} y\right)}\right]$$

when

$$\delta y\big|_{\partial D} = \partial_m \delta y\big|_{\partial D} = \cdots = \partial_{m_1}\partial_{m_2} \cdots \partial_{m_{s-1}} \delta y\big|_{\partial D} = 0 .$$

1.18. Derive the Euler equations for the following lagrangians:

i. $a(x^k)\, y^2 + \log\left\{1 + \left[a^m(x^k)\, \partial_m y\right]^2\right\}$,

ii. $y^2 + a^{km} \partial_k y\, \partial_m y + \sin\left[k(x^k; y)^2\right]$ where

$$k(x^k; y) = \int_{D^*} K(x^k, z^k)\, y(z^k)\, dV(z),$$

iii. $\cosh\left[y a^m \partial_m y + k(x^k; y) - 7 b\, k(x^k; y)^2\right]$ where

$$k(x^k; y) = \int_{D^*} H^m(x^k, z^k)\, \partial_m y(z^k)\, dV(z).$$

1.19. Let $a^{km}(x) = a^{mk}(x)$ be positive definite and nonsingular for all $x$ in $D^*$, and set $L = a^{km}\partial_k y \partial_m y$. Obtain and interpret the natural boundary conditions for this problem ((1.118) or $\delta y|_{\partial D} = 0$). Find the most general form of $L^m$ such that the boundary conditions

$$N_m(x)\left[a^{mk}\partial_k y + f^m(x)\right]\Big|_{\partial D} = [u(x)\,y]\Big|_{\partial D}$$

can be obtained.

1.20. Find the natural boundary conditions for Cases 5, 6, 8 and 9 in Sec. 1.7.

# 2 Variational Calculus for Several Dependent Functions

## 2.1. THE NORM FOR PRODUCTS OF FUNCTION SPACES

Chapter 1 was concerned with problems involving only one dependent function. It was sufficient in these cases to consider the linear spaces of functions of class $C$ or of class $C^1$ with the norm

$$\|y\| = \max_{x \in D^*} |y(x^k)|,$$

in which case the space was denoted by $\mathcal{C}(D^*)$, or with the norm

$$\|y\| = \max_{x \in D^*} |y(x^k)| + \sum_{k=1}^{n} \max_{x \in D^*} |\partial_k y(x^m)|,$$

in which case the space was denoted by $\mathcal{D}_1(D^*)$. In this chapter we will extend our study to problems involving two or more dependent functions. To consider ordered collections of such functions, we must extend the norm to a multiple Cartesian product of function spaces.

Let $\{\Phi_\Lambda(x^k)\}$, $\Lambda = 1, \ldots, N$, denote an ordered collection of $N$ functions of class $C$ or of class $C^1$ on $D^*$. With the operations of addition and scalar multiplication defined by

$$\{\Phi_\Lambda + \Psi_\Lambda\}(x^k) = \{\Phi_\Lambda(x^k)\} + \{\Psi_\Lambda(x^k)\}, \tag{2.1}$$

$$a \cdot \{\Phi_\Lambda(x^k)\} = \{a\Phi_\Lambda(x^k)\}, \tag{2.2}$$

the collection of all such $N$-tuples forms a linear space. If each of the $\{\Phi_\Lambda(x^k)\}$ is of class $C$, we define the norm by

$$\|\{\Phi_\Lambda\}\| = \sum_{\Lambda=1}^{N} \max_{x \in D^*} |\Phi_\Lambda(x^k)| \tag{2.3a}$$

or by

$$\|\{\Phi_\Lambda\}\| = \sqrt{\sum_{\Lambda=1}^{N}\left(\max_{x \in D^*} |\Phi_\Lambda(x^k)|\right)^2} . \tag{2.3b}$$

The space with either of these norms is denoted by $\mathcal{C}(D^*;N)$ and is essentially a vector space of function spaces. This is most evident for the norm given by (2.3b). If each of the $\Phi_\Lambda(x^k)$ is of class $C^1$ on $D^*$, the norm is defined by

$$\|\{\Phi_\Lambda\}\| = \sum_{\Lambda=1}^{N}\left(\max_{x \in D^*} |\Phi_\Lambda(x^k)| + \sum_{m=1}^{n} \max_{x \in D^*} |\partial_m \Phi_\Lambda(x^k)|\right) \tag{2.4a}$$

or by

$$\|\{\Phi_\Lambda\}\| = \sqrt{\sum_{\Lambda=1}^{N}\left(\max_{x \in D^*} |\Phi_\Lambda(x^k)| + \sum_{m=1}^{n} \max_{x \in D^*} |\partial_m \Phi_\Lambda(x^k)|\right)^2} . \tag{2.4b}$$

A space with either of these norms is denoted by $\mathfrak{D}_1(D^*;N)$, and a vector space of function spaces is again obtained. In all cases, we have

$$\mathcal{C}(D^*;1) \equiv \mathcal{C}(D^*), \qquad \mathfrak{D}_1(D^*;1) \equiv \mathfrak{D}_1(D^*). \tag{2.5}$$

## 2.2. STATIONARIZATION AND THE EULER-LAGRANGE OPERATORS

Let $g_a[x^k, z^k, \Phi_\Lambda(z^k), \partial_m \Phi_\Lambda(z^k)]$ be $q$ functions of class $C^2$ in their $2n + N + nN$ arguments. To simplify the writing of these and similar functions, we introduce the notation ${}_\alpha\Phi_\Lambda(x^k)$, in which $\alpha$ runs from zero through $n$ and

$${}_0\Phi_\Lambda(x^k) = \Phi_\Lambda(x^k), \qquad {}_m\Phi_\Lambda(x^k) = \partial_m \Phi_\Lambda(x^k). \tag{2.6}$$

For each point $\{x^k\}$ in $D^*$, we define $q$ functionals by the relations

$$k_a(x^k;\{\Phi_\Lambda\}) = \int_{D^*} g_a[x^k, z^k, {}_\alpha\Phi_\Lambda(z^k)]\, dV(z). \tag{2.7}$$

Let $L[x^k, {}_\alpha\Phi_\Lambda(x^k), k_a(x^k; \{\Phi_\Lambda\})]$ be a function of class $C^2$ in its $n + N(n+1) + q$ arguments and define the functional $J[\{\Phi_\Lambda\}](L)$ by the relation

$$J[\{\Phi_\Lambda\}](L) = \int_{D^*} L[x^k, {}_\alpha\Phi_\Lambda(x^k), k_a(x^k; \{\Phi_\Lambda\})]\, dV(x). \tag{2.8}$$

Continuity, differentiability and linearity of functionals of several functions are defined in the same manner as for functionals of a single function, with the exception that the norm of $\mathfrak{D}_1(D^*; N)$ is used instead of the norm of $\mathfrak{D}_1(D^*)$.

An element of $\mathfrak{D}_1(D^*; N)$ is said to *stationarize* $J[\{\Phi_\Lambda\}](L)$ if and only if

$$\delta J[\{\Phi_\Lambda\}](L) = 0 \tag{2.9}$$

for all variations of $\{\Phi_\Lambda(x^m)\}$ that vanish on the boundary of $D^*$. Here, $\delta J[\{\Phi_\Lambda\}](L)$ is defined in the same was in Chap. 1: if

$$J[\{\Phi_\Lambda + h_\Lambda\}](L) - J[\{\Phi_\Lambda\}](L) = j[\{\Phi_\Lambda\}, \{h_\Lambda\}](L) + o(\|\{h_\Lambda\}\|),$$

where $j[\{\Phi_\Lambda\}, \{h_\Lambda\}](L)$ is a linear functional of $\{h_\Lambda\}$ for each fixed $\{\Phi_\Lambda\}$, then $j[\{\Phi_\Lambda\}, \{h_\Lambda\}](L) = \delta J[\{\Phi_\Lambda\}](L)$. The collection of all elements of $\mathfrak{D}_1(D^*; N)$ that stationarize $J[\{\Phi_\Lambda\}](L)$ for a given $L$ is denoted by $S(L)$ and is referred to as the *stationarization set base L*.

Set

$$\Delta J[\{\Phi_\Lambda\}, \{h_\Lambda\}](L) = J[\{\Phi_\Lambda + h_\Lambda\}](L) - J[\{\Phi_\Lambda\}](L), \qquad \{h_\Lambda\} = \{\delta\Phi_\Lambda\}; \tag{2.10}$$

then

$$\Delta J[\{\Phi_\Lambda\}, \{h_\Lambda\}](L) = \int_{D^*} \left[ \frac{\partial L}{\partial \Phi_\Lambda}\, \delta\Phi_\Lambda + \frac{\partial L}{\partial(\partial_m \Phi_\Lambda)}\, \partial_m \delta\Phi_\Lambda + \frac{\partial L}{\partial k_a}\, \delta k_a \right] dV(x) \tag{2.11}$$
$$+ \, o(\|\{\delta\Phi_\Lambda\}\|),$$

where the quantity $\delta k_a$ is computed in exactly the same way as in Chap. 1. This result allows computation of $\delta J[\{\Phi_\Lambda\}](L)$. If $\delta J[\{\Phi_\Lambda\}](L)$ is to vanish for all variations $\{\delta\Phi_\Lambda(x^k)\}$ that vanish on $\partial D^*$, it must vanish for the $N$ independent variations such that only one of the $N$ functions $\delta\Phi_\Lambda(x^k)$ is nonzero in each (i.e., it must vanish for each element of a basis for the "vector" of variations). Thus the problem with $N$ functions is converted into $N$ problems with one function with the remaining functions as parameters. A direct application of the results obtained in Chap. 1 leads to Theorem 2.1 when caution is exercised about the existence of second derivatives.

**Theorem 2.1.** *Let* $L$ *and* $g_a$ *be functions of class* $C^2$ *in their respective arguments and let* $\{\Phi_\Lambda(x^k)\}$ *be functions of class* $C^2$ *at all points of* $D^*$ *except for those for which*

$$\frac{\partial^2 L}{\partial(\partial_n \Phi_\Sigma)\,\partial(\partial_m \Phi_\Lambda)} + \int_{D^*} \frac{\partial L}{\partial k_a}(z) \frac{\partial^2 g_a^*}{\partial(\partial_n \Phi_\Sigma)\,\partial(\partial_m \Phi_\Lambda)}\, dV(z) = 0. \quad (2.12)$$

*Then* $\{\Phi_\Lambda(x^k)\} \in S(L)$ *if and only if the* $N$ *functions* $\Phi_\Lambda(x^k)$ *satisfy the* $N$ *Euler equations*

$$\{\mathcal{E} \mid L\}_{\Phi_\Lambda}(x^m) = 0, \quad (2.13)$$

*where*

$$\{\mathcal{E} \mid L\}_{\Phi_\Lambda}(x^k) = \{e \mid L(k_a)\}_{\Phi_\Lambda}(x^k) + \int_{D^*} \frac{\partial L}{\partial k_a}(z^k)\{e \mid g_a^*(z^k)\}_{\Phi_\Lambda}(x^k)\, dV(z) \quad (2.14)$$

*is the (nonlocal) Euler-Lagrange operator with respect to* $\Phi_\Lambda$ *and*

$$\{e \mid L\}_{\Phi_\Lambda}(x^k) = \frac{\partial L}{\partial \Phi_\Lambda} - \partial_m \left[ \frac{\partial L}{\partial(\partial_m \Phi_\Lambda)} \right] \quad (2.15)$$

*is the little Euler-Lagrange operator with respect to* $\Phi_\Lambda$.

The notation in the above equations is the same as that introduced in Chap. 1: $(\partial L/\partial k_a)(z^k)$ denotes the quantity that is obtained by differentiating $L$ with respect to $k_a$ and then replace $x^k$ by $z^k$ throughout; $g_a^*$ is defined by

$$g_a = g_a[x^k, z^k, {}_\alpha\Phi_\Lambda(z^k)], \qquad g_a^* = g_a[z^k, x^k, {}_\alpha\Phi_\Lambda(x^k)].$$

Stationarization in the case of several functions with surface integrals and without requirement concerning the vanishing of variations on the boundary of $D^*$ is left as an exercise. The student is advised to follow the results of Sec. 1.9.

As an example, set

$$k(x^k; \Phi_1, \Phi_2) = \int_{D^*} K(x^k, z^k)\,\Phi_1(z^k)\,dV(z) + \int_{D^*} H(x^k, z^k)\,\Phi_2(z^k)\,dV(z)$$

and $L = a^{km}\partial_k \Phi_1 \partial_m \Phi_2 + \Phi_1 k(x^k; \Phi_1, \Phi_2)$. Then (2.13) through (2.15) give

$$\{\mathcal{E}|L\}_{\Phi_1}(x^k) = -\partial_k(a^{km}\partial_m \Phi_2) + k(x^k; \Phi_1, \Phi_2) + \int_{D^*} \Phi_1(z^k)\,K(z^k, x^k)\,dV(z)\,,$$

$$\{\mathcal{E}|L\}_{\Phi_2}(x^k) = -\partial_m(a^{km}\partial_k \Phi_1) + \int_{D^*} \Phi_2(z^k)\,H(z^k, x^k)\,dV(z)\,.$$

## 2.3. THE NULL CLASS OF THE EULER-LAGRANGE OPERATORS

At this point it is clear that when $N$ functions are involved in the lagrangian function, a system of $N$ Euler equations is obtained: stationarization over an "$N$-vector of function spaces" results in an "$N$-vector" of Euler equations. Thus the results of Chap. 1 apply to the present discussion if allowance is made for the $N$-vector structure. The form of these results is evident, although the details are in some instances more complicated than in the case of $\mathfrak{D}_1(D^*)$. A case in point is the null class of the Euler-Lagrange operators.

It should also be noted that we must deal with a more constrained continuity structure than in the case of only one dependent function. This added constraint is a result of having to deal with the commutativity structure involving the $N$ functions as well as with the $n$ independent variables. In most cases it is sufficient to restrict our considerations to the class of $C^2$ functions contained in $\mathfrak{D}_1(D^*; N)$. The additional restrictions which must be made in order to obtain results for arbitrary element of $\mathfrak{D}_1(D^*)$ will be indicated in the following discussion.

A lagrangian function is said to be an element of $\mathfrak{N}(N)$, *the null class of the Euler-Lagrange operators*, if and only if

$$\{\mathcal{E}|L\}_{\Phi_\Lambda}(x^k) = 0\,, \qquad \Lambda = 1, \ldots, N\,, \tag{2.16}$$

holds for every $N$-vector of $C^2$ functions belonging to $\mathfrak{D}_1(D^*; N)$. The *trivial null class*, $\mathfrak{N}(N)^*$, consists of all lagrangian functions for which $J[\{\Phi_\Lambda\}](L)$ is constant in value as a function on $\mathfrak{D}_1(D^*; N)$.

**Theorem 2.2.** *The most general function,* $A[x^k, {}_{\alpha}\Phi\,(x^k)]$, *that is of class* $C^2$ *in its* $n + N(n+1)$ *arguments and is such that*

$$\{e\,|\,A\}_{\Phi_\Lambda}(x^k) = 0\,, \qquad \Lambda = 1, \ldots, N \tag{2.17}$$

*holds for every* N-*vector of* $C^2$ *functions contained in* $\mathfrak{D}_1(D^*; N)$, *is given by*

$$A = B + \sum_{s=1}^{r} s^{-1} P_s\,, \tag{2.18}$$

*where the following conditions are satisfied:*

i. $P_s = B^{\Lambda_1 k_1 \ldots \Lambda_s k_s} \partial_{k_1}\Phi_{\Lambda_1}\, \partial_{k_2}\Phi_{\Lambda_2} \cdots \partial_{k_s}\Phi_{\Lambda_s}.$ (2.19)

ii. *B are functions of* $x$ *and* $\Phi$, *of class* $C^2$ *and, for* $s > 1$, *are completey skey symmetric in the Latin indices and in the Greek indices.*

iii. $r = \min(n, N)\,.$ (2.20)

iv. $$0 = \partial^*_m B^{\Lambda m} - \frac{\partial B}{\partial \Phi}, \qquad \partial^*_m(\quad) = \partial_m(\quad)\Big|_{\Phi_\Lambda}, \tag{2.21}$$

$$0 = \frac{\partial B^{\Gamma i_1 \Lambda_2 i_2 \ldots \Lambda_s i_s}}{\partial \Phi_{\Lambda_i}} + \cdots + \frac{\partial B^{\Lambda_1 i_1 \ldots \Gamma i_s}}{\partial \Phi_{\Lambda_s}} - \frac{\partial B^{\Lambda_1 i_1 \ldots \Lambda_s i_s}}{\partial \Phi_\Gamma} + s\partial^*_m B^{m\bar{\Gamma}\Lambda_1 i_1 \ldots \Lambda_s i_s}, \tag{2.22}$$

$$0 = B^{\Gamma m \Lambda_1 i_1 \ldots \Lambda_r i_r}. \tag{2.23}$$

**Proof.** When (2.15) is used and the differentiation is carried out, (2.17) can be satisfied if and only if

$$\frac{\partial^2 A}{\partial(\partial_m\Phi_\Lambda)\,\partial(\partial_n\Phi_\Gamma)}\,\partial_m\partial_n\Phi_\Gamma + \frac{\partial^2 A}{\partial\Phi_\Gamma\,\partial(\partial_m\Phi_\Lambda)}\,\partial_m\Phi_\Gamma + \partial^*_m\frac{\partial A}{\partial(\partial_m\Phi_\Lambda)} - \frac{\partial A}{\partial\Phi_\Lambda} = 0 \tag{2.24}$$

holds for all $N$-vectors of $C^2$ functions contained in $\mathfrak{D}_1(D^*; N)$; that is, if and only if the coefficients of the second derivatives, the coefficients of the first derivatives, and the coefficients of $\Phi_\Lambda$ vanish separately. This can be the case if and only if

$$\frac{\partial^2 A}{\partial(\partial_m \Phi_\Lambda)\,\partial(\partial_n \Phi_\Gamma)}\,\partial_m \partial_n \Phi_\Gamma = 0 \tag{2.25}$$

is an identity for all $N$-vectors of $C^2$ functions in $\mathfrak{D}_1(D^*;N)$. Since we are dealing with $C^2$-structure, the order of second derivatives is commutative, and hence it is sufficient to require that

$$\frac{\partial^2 A}{\partial(\partial_m \Phi_\Lambda)\,\partial(\partial_n \Phi_\Gamma)} + \frac{\partial^2 A}{\partial(\partial_n \Phi_\Lambda)\,\partial(\partial_m \Phi_\Gamma)} = 0 \tag{2.26}$$

be an identity for all $N$-vectors of $C^2$ functions in $\mathfrak{D}_1(D^*;N)$. This system of equations is a system for the determination of $A$ (identical satisfaction with respect to $\Phi_\Lambda$ always leads to equations for the determination of $A$). Let $r$ be defined by (2.20), and consider the collection of functions

$$W^{\Lambda_1 i_1 \ldots \Lambda_{r+1} i_{r+1}} = \frac{\partial^{r+1} A}{\partial\left(\partial_{i_1}\Phi_{\Lambda_1}\right)\partial\left(\partial_{i_2}\Phi_{\Lambda_2}\right)\ldots\partial\left(\partial_{i_{r+1}}\Phi_{\Lambda_{r+1}}\right)}. \tag{2.27}$$

We wish to prove that $W$ vanishes identically as a consequence of (2.26). We obviously have two cases: $r = n$ and $r = N$. For the case $r = n$, at least two of the Latin indices are the same in each $W$, and hence each $W$ can be constructed from an appropriate $(r-1)$-fold differentiation of $K_{\Lambda\Gamma}$, where

$$K_{\Lambda\Gamma} = \not\sum_j \frac{\partial^2 A}{\partial(\partial_j \Phi_\Lambda)\,\partial(\partial_j \Phi_\Gamma)}$$

and $j$ is the index that occurs twice in the $W$ under consideration. Thus $K$ vanishes identically as a consequence of (2.26), and hence $W$ vanishes identically for $r = n$. For the case $r = N$, at least two of the Greek indices are the same for each $W$, and hence $W$ can be obtained as an appropriate $(r-1)$-fold differentiation of

$$K_{mn} = \not\sum_\Gamma \frac{\partial^2 A}{\partial(\partial_m \Phi_\Gamma)\,\partial(\partial_n \Phi_\Gamma)}.$$

Since $A$ is $C^2$, we have $K_{mn} = K_{nm}$, and accordingly, (2.26) implies the identical vanishing of $W$. Now, the identical vanishing of $W$ implies that $A$ is, at most, a polynomial of degree $r$ in the first partial derivatives of $\Phi$. (Note that the $W$ will not vanish unless there are both $r+1$ Greek indices and $r+1$ latin indices.) We are therefore led to the fact that $A$ must be of the form given by (2.18) where $P$ is defined by (2.19). The $s^{-1}$ is included in (2.18) for simplicity of notation. From the form of $P$, we may assume with no loss of generality that $B$ is such that

$$B^{\dots\Lambda_k i_k \dots \Lambda_t i_t \dots} = B^{\dots\Lambda_t i_t \dots \Lambda_k i_k \dots}, \tag{2.28}$$

and hence

$$\frac{\partial^2 P_s}{\partial(\partial_m \Phi_\Lambda)\,\partial(\partial_n \Phi_\Gamma)} = s(s-1)\,B^{\Lambda m \Gamma n \dots \Lambda_s i_s}\,\partial_{i_3}\Phi_{\Lambda_3} \cdots \partial_{i_s}\Phi_{\Lambda_s}. \tag{2.29}$$

Substituting (2.18) into (2.26), and making use of (2.29), we obtain

$$B^{\Lambda m \Gamma n \dots} + B^{\Lambda n \Gamma m \dots} = 0\,, \tag{2.30}$$

Thus, in order for (2.18) to satisfy (2.26), equations (2.28) and (2.30) must hold: we must require $B$ to be completely skew symmetric in the Latin indices and in the Greek indices. We have thus reduced our problem to that of establishing conditions on $B$ such that

$$\begin{aligned} 0 = -\frac{\partial B}{\partial \Phi_\Lambda} + \sum_{s=1}^{r} \Bigg\{ &\partial_m \Phi_\Gamma \frac{\partial B^{\Lambda m \dots \Lambda_s i_s}}{\partial \Phi_\Gamma}\,\partial_{i_2}\Phi_{\Lambda_2} \cdots \partial_{i_s}\Phi_{\Lambda_s} \\ &+ \partial^*_m B^{\Lambda m \dots \Lambda_s i_s}\,\partial_{i_2}\Phi_{\Lambda_2} \cdots \partial_{i_s}\Phi_{\Lambda_s} \\ &- \frac{1}{s}\frac{\partial B^{\Lambda_1 i_1 \dots \Lambda_s i_s}}{\partial \Phi_\Lambda}\,\partial_{i_1}\Phi_{\Lambda_1} \cdots \partial_{i_s}\Phi_{\Lambda_s} \Bigg\} \end{aligned} \tag{2.31}$$

is an identity over $\mathfrak{D}_1(D^*, N)$. Collecting the terms appropriately and equating the coefficients of the different powers of $\partial_m \Phi_\Lambda$ to zero, we obtain the system (2.21)-(2.23). This establishes Theorem 2.2.

The student is advised to check that Theorem 2.2, with $N = r = 1$, reduces to the corresponding result established in Sec. 1.6. If (2.17) is required to hold for all elements of $\mathfrak{D}_1(D^*; N)$, the Euler equations cannot involve second derivatives; that is, the lagrangian function must be linear in the derivatives of $\Phi$. Reducing Theorem 2.2 to this case, we see that the result is again given by (2.18), but the summation must terminate with $s = 1$. This case thus exactly reproduces the results of Chap. 1. All of the additional structure and the restriction to $C^2$ functions comes from the $N$-vector structure of the function space under consideration, namely, $\mathfrak{D}_1(D^*; N)$.

The system of equations (2.21)-(2.23) is of interest in its own right. If $r = 1$, we have

$$0 = \partial_m B^{\Lambda m} - \frac{\partial B}{\partial \Phi_\Lambda}, \qquad 0 = \frac{\partial B^{\Lambda m}}{\partial \Phi_\Sigma} - \frac{\partial B^{\Sigma m}}{\partial \Phi_\Lambda}, \tag{2.32}$$

which are necessary and sufficient conditions for the existence of a collection of functions $F^m[x^k, \Phi_\Lambda(x^k)]$ such that

$$B = \partial_m^* F^m, \qquad B^{\Lambda m} = \frac{\partial F^m}{\partial \Phi_\Lambda}. \tag{2.33}$$

For general $r$, if one starts with the first of equations (2.33) a lengthy but straightforward calculation shows that the system (2.21)-(2.23) is a necessary and sufficient condition for the existence of a collection of functions $F^{m\Lambda_i i_i \cdots}[x^k, \Phi_\Lambda(x^k)]$ which are completely skew symmetric in the Latin indices and in the Greek indices and are such that

$$B = \partial_m^* F^m, \tag{2.34}$$

$$B^{\Lambda m} = \frac{\partial F^m}{\partial \Phi_\Lambda} + \partial_t^* F^{t\Lambda m},$$

$$\vdots$$

$$B^{\Lambda_1 i_1 \ldots \Lambda_s i_s} = \frac{1}{(s-1)!}\left\{\frac{\partial F^{i_1 \Lambda_2 i_2 \ldots \Lambda_s i_s}}{\partial \Phi_{\Lambda_1}} + \frac{\partial F^{i_2 \Lambda_1 i_1 \ldots \Lambda_s i_s}}{\partial \Phi_{\Lambda_2}} + \cdots + \frac{\partial F^{i_s \Lambda_2 i_2 \ldots \Lambda_1 i_1}}{\partial \Phi_{\Lambda_s}} + \partial_t^* F^{t\Lambda_1 i_1 \ldots \Lambda_s i_s}\right\}, \tag{2.35}$$

$$F^{t\Lambda_1 i_1 \ldots \Lambda_r i_r} = 0. \tag{2.36}$$

When these results are combined with those of Theorem 2.2, we have the following generalization of the divergence content of $\mathfrak{N}(1)$ established in Chap. 1.

**Theorem 2.3.** *The most general lagrangian function that is annihilated by the little Euler-Lagrange operator is of the form*

$$\partial_m K^m(r),$$

*where*

$$K^m(r) = \sum_{s=0}^{r-1} \frac{M^m(s)}{s!} \tag{2.37}$$

$$M^n(s) = F^{n\Lambda_1 i_1 \ldots \Lambda_s i_s} \partial_{i_1} \Phi_{\Lambda_1} \ldots \partial_{i_s} \Phi_{\Lambda_s} \tag{2.38}$$

$$r = \min(n, N), \tag{2.39}$$

*and F are functions of class* $C^3$ *in their arguments* $[x^k, \Phi(x^k)]$, *which are completely skew symmetric in the Latin indices and in the Greek indices.*

Thus the case of several dependent functions gives more general divergence structures than those resulting in linear forms in the derivatives. However, it must be clearly recognized that a completely general divergence is still not variationally deletable.

Considerations similar to those of Chap. 1 now lead to the following general result.

**Theorem 2.4.** *The most general element of* $\mathfrak{N}(N)$ *is a sum of an element of the trivial null class,* $\mathfrak{N}^*(N)$, *and an element of the form*

$$\partial_m K^m(r),$$

*where*

$$K^m(r) = \sum_{s=0}^{r-1} \frac{M^m(s)}{s!} \tag{2.40}$$

$$M^n(s) = F^{n\Lambda_1 i_1 \ldots \Lambda_s i_s} \partial_{i_1} \Phi_{\Lambda_1} \ldots \partial_{i_s} \Phi_{\Lambda_s} \tag{2.41}$$

*are functions of class C in their* $n + N + q$ *arguments* $[x^k, \Phi_\Lambda(x^k), k_a(x^k; \{\Phi_\Lambda\})]$ *which are completely skew symmetric in the Latin indices and in the Greek indices;* $k_a(x^k; \{\Phi_\Lambda\})$ *are given by*

$$k_a(x^k; \{\Phi_\Lambda\}) = \int_{D^*} \frac{\partial}{\partial z^m} k_a{}^m(r)\, dV \tag{2.42}$$

*with*

$$k_a{}^m(r) = \sum_{s=0}^{r-1} \frac{m^m(s)}{s!}, \tag{2.43}$$

$$m_a{}^n(s) = f_a^{n\Lambda_1 i_1 \ldots \Lambda_s i_s} \partial_{i_1} \Phi_{\Lambda_1}(z) \ldots \partial_{i_s} \Phi_{\Lambda_s}(z), \tag{2.44}$$

*$f$ are functions of class $C^3$ in their $2n + N$ arguments $[x^k, z^k, \Phi(z^k)]$ which are completely skew symmetric in the Latin indices and in the Greek indices; and*

$$r = \min(n, N). \tag{2.45}$$

It should now be apparent how complicated the situation can become when $N$-vectors of function spaces must be considered. A further complication also present is that *there are lagrangian functions that are identically annihilated by* $\{\mathcal{E}|\ \}_{\Phi_1}$ *but not by* $\{\mathcal{E}|\}_{\Phi_\Sigma}$ when $\Sigma \neq 1$. A trival example is given by

$$L = \partial_m F^m(x^k, \Phi_1) + [\Phi_2(x^k)]^2.$$

Many cases could thus be listed, but to save time and work we will omit this procedure.

In the next section we shall need to know certain elements of $\mathfrak{N}(N)$ when the lagrangian function is allowed to depend on second derivatives of $\Phi_\Lambda$. Therefore Lemma 2.1 is stated with the proof left to the reader.

**Lemma 2.1.** *If the lagrangian function depends on second derivatives of $\Phi_\Lambda$ and if the variations and their first derivatives are assumed to vanish on the boundary of $D^*$, then*

$$\partial_m V^m[x^k, {}_\alpha\Phi_\Lambda(x^k)] \tag{2.46}$$

*is an element of $\mathfrak{N}(N)$ and*

$$\{e|L\}_{\Phi_\Lambda}(x^k) = \frac{\partial L}{\partial \Phi_\Lambda} - \partial_m\left[\frac{\partial L}{\partial(\partial_n \Phi_\Lambda)}\right] + \partial_n\partial_m\left[\frac{\partial L}{\partial(\partial_n\partial_m \Phi_\Lambda)}\right]. \tag{2.47}$$

Two lagrangian functions are equivalent if and only if there exists an element $A$ of $\mathfrak{N}(N)$ such that

$$L_1 = L_2 + A, \tag{2.48}$$

in which case we write

$$L_1 \overset{\Omega}{=} L_2 \tag{2.49}$$

The relation $\Omega$ is an equivalence relation and partitions the collection of all lagrangian functions into equivalence classes. We denote the equivalence class containing a lagrangian function $L$ by $\hat{L}$. This many-to-one association of lagrangian functions and Euler equations will add complications to what follows.

An additional aspect of null classes which enteres into several arguments essential to our considerations must be noted. If an element of the null class contains functionals as arguments, these functional arguments are such that they can be reduced to surface integrals. This leads to situations in which a given lagrangian function with arguments in $\mathfrak{D}_1(D^*;N)$ can be an element of $\mathfrak{N}(p)$ for $p < N$, but not an element of $\mathfrak{N}(N)$. For example, if $N = 2$ and

$$L = \partial_m P^m, \qquad P^m = \int_{D^*} \Phi_1(x^k)\Phi_2(z^k) H^m(x^k, z^k)\, dV(z), \tag{2.50}$$

then

$$\{\mathcal{E} | L\}_{\Phi_1} \equiv 0, \qquad \{\mathcal{E} | L\}_{\Phi_2} = \int_{D^*} \frac{\partial}{\partial z^m} [\Phi_1(z^k) H^m(z^k, x^k)]\, dV(z), \tag{2.51}$$

and hence $L \in \mathfrak{N}(1)$, but $L \notin \mathfrak{N}(2)$ in general. The above situation results because $L$ has a divergence structure with respect to $\Phi_1(x^k)$, but not a divergence structure with respect to $\Phi_2(x^k)$ and because lagrangian functions which have functionals as arguments are considered. The situation is not usually treated in the literature and arises only when there are integrodifferential equations as Euler equations; if the Euler equations are differential equations, the above situation does not arise.

## 2.4. VARIATIONAL EMBEDDING, FUNDAMENTAL ADJOINT THEOREMS, AND EXISTING CONDITIONS

In order to study linear operators and linear Euler equations, the following definition is required. A linear integrodifferential operator $\mathcal{O}$ is said to be of *second order* if and only if the evaluation of $\mathcal{O}$, applied to any function in its domain, requires the calculation of derivatives of order less than three and integrals of derivatives of order less than two.

In Chap. 1 it was established that any linear integrodifferential operator of second order, whose domain consisted of 1-tuples of functions, could be embedded in a variational statement ($\mathcal{O}$ was the Euler-Lagrange operator for some lagrangian function) if and only if $\mathcal{O}$ was self-adjoint. The freedom afforded by considering $N$-vectors of functions provides the added generality for establishing the embedding of any linear integrodifferential operator in a variational statement. A system of basic adjoint theorems by which this is established will be proved in the subsequent considerations.

First, recall the definition of the adjoint operator, $\mathcal{O}^+$, of a given linear operator $\mathcal{O}$:

$$\int_{D^*} \{\psi \mathcal{O} \phi - \phi \mathcal{O}^+ \psi\} dV(x) = \int_{D^*} \partial_m P^m \, dV(x) = \int_{\partial D} P^m dS_m(x) .$$

This is the customary definition in that both $\mathcal{O}$ and $\mathcal{O}^+$ are restricted to being linear operators. However, this restriction is unnecessary since it is the *homogeneous* aspect rather than the linear aspect which is important.

Let $\mathcal{O}$ be a differential operator of second order which is *homogeneous of degree* $n \neq 0$; that is, the following identity holds for all $\phi(x^k)$ of class $C^2$:

$$n\mathcal{O}\phi \equiv \phi \frac{\partial(\mathcal{O}\phi)}{\partial\phi} + \frac{\partial(\mathcal{O}\phi)}{\partial(\partial_m\phi)} \partial_m\phi + \frac{\partial(\mathcal{O}\phi)}{\partial(\partial_m\partial_k\phi)} \partial_m\partial_k\phi .$$

This definition is easily extended to integrodifferential operators. Such an operator is said to be *homogeneous of degree* $n \neq 0$ if and only if the following identity holds.

$$n\mathcal{O}\phi \equiv \phi \frac{\partial(\mathcal{O}\phi)}{\partial\phi} + \frac{\partial(\mathcal{O}\phi)}{\partial(\partial_m\phi)} \partial_m\phi + \frac{\partial(\mathcal{O}\phi)}{\partial(\partial_m\partial_k\phi)} \partial_m\partial_k\phi$$
$$+ \int_{D^*} \frac{\partial(\mathcal{O}\phi)}{\partial k_a} \left\{ \frac{\partial g_a}{\partial\phi(z)} \phi(z) + \frac{\partial g_a}{\partial[\partial_m\phi(z)]} \partial_m\phi(z) \right\} dV(z) .$$

Here, the $k_a$ comprises all integrals of $\phi(z^k)$ and its first derivatives which occur in $\mathcal{O}\phi$ and $k_a = \int_{D^*} g_a[x^k, z^k, \phi(z^k), \partial_m\phi(z^k)] dV(z)$. For simplicity, the discussion is restricted to differential operators; the case of integrodifferential operators is left as an exercise for the student.

If $a$ = constant, then

$$\phi\{\mathcal{E}\,|\,a\psi\mathcal{O}\phi\}_\phi = a\phi\left\{\psi\frac{\partial(\mathcal{O}\phi)}{\partial\phi} - \partial_m\left[\psi\frac{\partial(\mathcal{O}\phi)}{\partial(\partial_m\phi)}\right] + \partial_m\partial_t\left[\psi\frac{\partial(\mathcal{O}\phi)}{\partial(\partial_m\partial_t\phi)}\right]\right\}.$$

The term $\phi[\partial(\mathcal{O}\phi)/\partial\phi]$ can now be solved for from the defining equation of homogeneity of degree $n$. After computation, we have

$$\phi\{\mathcal{E}\,|\,a\psi\mathcal{O}\phi\}_\phi = a\left\{n\psi\mathcal{O}\phi - \partial_m\left[\psi\phi\frac{\partial(\mathcal{O}\phi)}{\partial(\partial_m\phi)}\right.\right.$$

$$\left.\left. + \phi\partial_t\left(\psi\frac{\partial(\mathcal{O}\phi)}{\partial(\partial_m\partial_t\phi)}\right) - \psi(\partial_t\phi)\frac{\partial(\mathcal{O}\phi)}{\partial(\partial_m\partial_t\phi)}\right]\right\}.$$

Accordingly, if we set $a = n^{-1}$ and use the definition of the adjoint, then

$$\left\{\mathcal{E}\,\middle|\,\frac{1}{n}\psi\mathcal{O}\phi\right\}_\phi = \mathcal{O}^+\psi$$

and

$$P^m = \frac{1}{n}\left\{\psi\phi\frac{\partial(\mathcal{O}\phi)}{\partial(\partial_m\phi)} + \phi\partial_t\left[\psi\frac{\partial(\mathcal{O}\phi)}{\partial(\partial_m\partial_t\phi)}\right] - (\partial_t\phi)\frac{\partial(\mathcal{O}\phi)}{\partial(\partial_m\partial_t\phi)}\psi\right\}.$$

In other words,

$$\psi\mathcal{O}\phi - \phi\mathcal{O}^+\psi = \partial_m P^m;$$

and, hence,

$$\int_{D^*}\{\psi\mathcal{O}\phi - \phi\mathcal{O}^+\psi\}dV(x) = \int_{D^*}\partial_m P^m\,dV(x).$$

As an example, let

$$\mathcal{O}\phi = f^{mn}\partial_m\phi\partial_n\phi, \qquad f^{mn} = f^{nm}.$$

Then we have $n = 2$,

$$\mathcal{O}^+\psi = -\partial_m(\psi f^{mt}\partial_t\phi),$$

$$P^m = \psi\phi f^{mt}\partial_t\phi.$$

If

$$\mathcal{O}\phi = \sqrt{(\lambda^2\phi^2 + f^{mt}\partial_m\phi\partial_t\phi)} + \eta\phi, \qquad f^{mn} = f^{nm},$$

then $n = 1$,

$$\mathcal{O}^+\psi = \psi\left[\eta + \frac{\lambda^2\phi}{\sqrt{(\lambda^2\phi^2 + f^{mt}\partial_m\phi\partial_t\phi)}}\right] - \partial_m\left[\frac{\psi f^{mt}\partial_t\phi}{\sqrt{(\lambda^2\phi^2 + f^{ks}\partial_k\phi\partial_s\phi)}}\right],$$

$$P^m = \frac{\psi\phi f^{mt}\partial_t\phi}{\sqrt{(\lambda^2\phi^2 + f^{ks}\partial_k\phi\partial_s\phi)}}.$$

The preceding calculation shows that the Euler-Lagrange operator gives an automatic adjoint calculator for homogeneous differential operators of the second order. That this is also the case for any linear, second order, integrodifferential operator will be established in Theorem 2.5.

**Theorem 2.5.** *Let $\mathcal{O}$ denote any linear integrodifferential operator of second order, and let $\mathcal{O}^+$ denote the adjoint of $\mathcal{O}$. We have*

$$\{\mathcal{E}\,|\,\Psi\mathcal{O}\Phi\}_\Phi(x^k) = \mathcal{O}^+\Psi \tag{2.52a}$$

$$\{\mathcal{E}\,|\,\Psi\mathcal{O}\Phi\}_\Psi(x^k) = \mathcal{O}\Phi \tag{2.52b}$$

*and*

$$\{\mathcal{E}\,|\,\Phi\mathcal{O}^+\Psi + \partial_m P^m\}_\Psi = \mathcal{O}\Phi \tag{2.53a}$$

$$\{\mathcal{E}\,|\,\Phi\mathcal{O}^+\Psi + \partial_m P^m\}_\Phi = \mathcal{O}^+\Psi, \tag{2.53b}$$

*(i.e., the Euler-Lagrange operator is an automatic adjoint calculator).*

**Proof.** Any linear integrodifferential operator $\mathcal{O}$ of second order is necessarily of the form

$$\begin{aligned}\mathcal{O}\Phi(x) &= a(x^m)\Phi(x^m) + a^k(x^m)\partial_k\Phi(x^m) + a^{km}(x^m)\partial_k\partial_m\Phi(x^m)\\ &+ \int_{D^*} K(x^m, z^m)\Phi(z^m)\,dV(z)\\ &+ \int_{D^*} H^k(x^m, z^m)\partial_k\Phi(z^m)\,dV(z)\ ,\end{aligned}\tag{2.54}$$

with $a^{km} = a^{mk}$. It adjoint is given by

$$\begin{aligned}\mathcal{O}^+\Psi(x^m) &= a(x^m)\Psi(x^m) - \partial_k[a^k(x^m)\Psi(x^m)] + \partial_k\partial_m[a^{km}(x^m)\Psi(x^m)]\\ &+ \int_{D^*} K(z^m, x^m)\Psi(z^m)\,dV(z)\\ &- \int_{D^*} \Psi(z^m)\frac{\partial K^k(z^m, x^m)}{\partial x^k}\,dV(z)\ .\end{aligned}\tag{2.55}$$

Since $\{\mathcal{E}\,|\,L_1 + L_2\}_{\Phi_\Lambda} = \{\mathcal{E}\,|\,L_1\}_{\Phi_\Lambda} + \{\mathcal{E}\,|\,L_2\}_{\Phi_\Lambda}$, a direct substitution of each of the terms occurring on the right-hand side of (2.54) into (2.52) leads to the corresponding term on the right-hand side of (2.55), and conversely.

An alternative proof of (2.52) and a proof of (2.53) are given below. A basic problem in these proofs is the necessity of considering integral as well as differential operators, which leads to two fundamentally different cases.

*Case 1.* In this case we have the local relation,

$$\Psi\mathcal{O}\Phi - \Phi\mathcal{O}^+\Psi = \partial_m P^m\,,$$

in which the $n$ functions $\{P^m\}$ are functions of the $x$'s, $\Phi(x^k)$, $\Psi(x^k)$, the derivatives of these two functions, and are such that each $P^m$ is bilinear in $\Phi$ and $\Psi$. When (2.54) and (2.55) are used, it is easily seen that this case obtains only when $\mathcal{O}$ involves differential operations. Accordingly, since we have the situation in which the lagrangian function depends on second derivatives, Lemma 2.1 shows that $\partial_m P^m$ belongs to the null class $\mathfrak{N}(2)$. Therefore we have

$$0 \equiv \{\mathcal{E}\,|\,\partial_m P^m\}_\Phi = \{\mathcal{E}\,|\,\Psi\mathcal{O}\Phi - \Phi\mathcal{O}^+\Psi\}_\Phi = \{\mathcal{E}\,|\,\Psi\mathcal{O}\Phi\}_\Phi - \mathcal{O}^+\Psi,$$

$$0 \equiv \{\mathcal{E}\,|\,\partial_m P^m\}_\Psi = \{\mathcal{E}\,|\,\Psi\mathcal{O}\Phi - \Phi\mathcal{O}^+\Psi\}_\Psi = \mathcal{O}\Phi - \{\mathcal{E}\,|\,\Phi\mathcal{O}^+\Psi\}_\Psi.$$

*Case 2.* In this case we have the nonlocal situation $\Psi\mathcal{O}\Phi - \Phi\mathcal{O}^{+}\Psi \neq \partial_m P^m(x^k, \Psi, \Phi, \ldots)$ but

$$\int_{D^*} \{\Psi\mathcal{O}\Phi - \Phi\mathcal{O}^{+}\Psi\} dV(x) = \int_{D^*} \partial_m P^m dV(x) .$$

In the above situation $\partial_m P^m$ is annihilated by $\{\mathcal{E}|\ \}_\Phi$ but not by $\{\mathcal{E}|\ \}_\Psi$. To understand what occurs, consider the situation in which

$$\mathcal{O}\Phi = \int_{D^*} H^m(x^k, z^k) \frac{\partial\Phi(z^k)}{\partial z^m} dV(z)$$

$$\mathcal{O}^{+}\Psi = -\int_{D^*} \Psi(z^k) \frac{\partial H^m(z^k, x^k)}{\partial x^m} dV(z) .$$

We then have

$$\Psi\mathcal{O}\Phi - \Phi\mathcal{O}^{+}\Psi = \int_{D^*} \left[ \Psi(x) H^m(x, z) \frac{\partial\Phi(z)}{\partial z^m} + \Phi(x)\Psi(z) \frac{\partial H^m(z, x)}{\partial x^m} \right] dV(z)$$

However,

$$\neq \partial_m P^m.$$

$$\int_{D^*} \{\Psi\mathcal{O}\Phi - \Phi\mathcal{O}^{+}\Psi\} dV(x) = \int_{D^*}\int_{D^*} \left[ \Psi(x) H^m(x, z) \frac{\partial\Phi(z)}{\partial z^m} \right.$$

$$\left. + \Phi(x)\Psi(z) \frac{\partial H^m(z, x)}{\partial x^m} \right] dV(z)$$

$$= \int_{D^*} \frac{\partial}{\partial x^m} \left[ \Phi(x) \int_{D^*} \Psi(z) H^m(z, x) dV(z) \right] dV(x)$$

so that $P^m = \Phi(x)\int_{D^*} \Psi(z) H^m(z, x) dV(z)$ . We accordingly obtain $\left\{\mathcal{E} | \partial_m P^m\right\}_\Phi \equiv 0$, but

$$\left\{\mathcal{E} | \partial_m P^m\right\}_\Psi = \int_{D^*} \left[ H^m(x, z) \frac{\partial\Phi(z)}{\partial z^k} + \Phi(z) \frac{\partial H^k(x, z)}{\partial z^k} \right] dV(z)$$

$$= \int_{D^*} \frac{\partial}{\partial z^k} \left( H^k(x, z)\Phi(z) \right) dV(z) \neq 0 .$$

Consider the quantity $T(\Psi, \Phi)$ defined by

$$T = \Psi \mathcal{O} \Phi - \Phi \{\mathcal{E} \mid \Psi \mathcal{O} \Phi\}_{\Phi}.$$

Since $\mathcal{O}$ is a linear operator, there exists a linear operation $\hat{\mathcal{O}}$ such that

$$\hat{\mathcal{O}} \Psi = \{\mathcal{E} \mid \Psi \mathcal{O} \Phi\}_{\Phi},$$

and hence $T = \Psi \mathcal{O} \Phi - \Phi \hat{\mathcal{O}} \Psi$. Thus, $T(\Psi, \Phi)$ is bilinear (homogeneous) in $(\Psi, \Phi)$. Now

$$\{\mathcal{E} \mid T\}_{\Phi} = \{\mathcal{E} \mid \Psi \mathcal{O} \Phi - \Phi \mathcal{O} \Psi\}_{\Phi} = \{\mathcal{E} \mid \Psi \mathcal{O} \Phi\}_{\Phi} - \hat{\mathcal{O}} \Psi = \hat{\mathcal{O}} \Psi - \hat{\mathcal{O}} \Psi \equiv 0,$$

and, accordingly, $T$ is an element of $\mathfrak{N}(1)$ with respect to $\Phi(x^k)$. The results of Chap. 1 concerning the characterization of $\mathfrak{N} = \mathfrak{N}(1)$ can be used, and we accordingly conclude that there exists an element

$$A[x^k; \Phi(x^k)] = a[x^k, \Psi(x^k); \Phi(x^k)]$$

of $\mathfrak{N}^*(1)$ (with respect to $\Phi(x^k)$) and a collection of $n$ functions

$$P^m[x^k; \Phi(x^k)] = p^m[x^k, \Psi(x^k); \Phi(x^k)]$$

such that

$$T = \partial_m P^m[x^k; \Phi(x^k)] + A[x^k; \Phi(x^k)].$$

Since $T$ is bilinear (homogeneous) in $(\Psi, \Phi)$, the $n+1$ functions $a[x^k, \Psi(x^k); \Phi(x^k)]$ and $p^m[x^k, \Psi(x^k); \Phi(x^k)]$ are clearly bilinear (homogeneous) in $(\Phi, \Psi)$. Since $A \in \mathfrak{N}^*(1)$, we have that

$$\int_{D^*} A[x^k; \Phi(x^k)]\, dV(x) = \int_{D^*} a[x^k, \Psi(x^k); \Phi(x^k)]\, dV(x)$$

is constant in value for each fixed $\Psi(x^k)$ and all $\Phi(x^k)$. Therefore,

$$\int_{D^*} A[x^k; \Phi(x^k)]\, dV(x) = M[\Psi] = \int_{D^*} a[x^k, \Psi(x^k); \Phi(x^k)]\, dV(x)$$

where $M$ is a functional of $\Psi(x^k)$ only. Using this independence of $M[\Psi]$ on the choice of $\Phi(x^k)$, we must accordingly have

$$M[\Psi] = \int_{D^*} a[x^k, \Psi(x^k); 0]\, dV(x).$$

The bilinear (homogeneous) structure of $a[x^k, \Psi(x^k); \Phi(x^k)]$ then gives

$$a[x^k, \Psi(x^k); 0] = 0,$$

from which we conclude that $M[\Psi] = 0$ for all $\Psi(x^k)$. We have thus established that $a[x^k, \Psi(x^k); \Phi(x^k)]$ is actually an element of $\mathfrak{N}^*(2)$ and that

$$\int_{D^*} A[x^k; \Phi(x^k)]\,dV(x) = \int_{D^*} a[x^k, \Psi(x^k); \Phi(x^k)]\,dV(x) = 0.$$

When the two forms of $T$ are combined, we have

$$\Psi\mathcal{O}\Phi - \Phi\hat{\mathcal{O}}\Psi = \partial_m P^m + A.$$

An integration of this equality over D* then gives

$$\int_{D^*} \{\Psi\mathcal{O}\Phi - \Phi\hat{\mathcal{O}}\Psi\}dV(x) = \int_{D^*} \{\partial_m P^m + A\}dV(x) = \int_{D^*} \partial_m P^m dV(x),$$

and hence

$$\mathcal{O}^+\Psi = \hat{\mathcal{O}}\Psi = \{\mathcal{E}\,|\,\Psi\mathcal{O}\Phi\}_\Phi.$$

We thus see that we are indeed in the case where

$$\Psi\mathcal{O}\Phi - \Phi\hat{\mathcal{O}}\Psi \neq \partial_m P^m.$$

We now have

$$\Psi\mathcal{O}\Phi - \Phi\mathcal{O}^+\Psi = \partial_m P^m + A;$$

and, accordingly,

$$\left\{\mathcal{E}\,|\,\Phi\mathcal{O}^+\Psi + \partial_m P^m\right\}_\Psi = \{\mathcal{E}\,|\,\Psi\mathcal{O}\Phi - A\}_\Psi = \mathcal{O}\Phi, \qquad [A \in \mathfrak{N}^*(z)].$$

Thus, Theorem 2.5 is established. In general, $\left\{\mathcal{E}\,|\,\partial_m P^m\right\}_\Psi \neq 0$, although we always have $\left\{\mathcal{E}\,|\,\partial_m P^m\right\}_\Phi = 0$; in fact, we have $\{\mathcal{E}\,|\,T\}_\Psi = \left\{\mathcal{E}\,|\,\partial_m P^m\right\}_\Psi$. If we had both $\left\{\mathcal{E}\,|\,\partial_m P^m\right\}_\Psi = 0$ and $\left\{\mathcal{E}\,|\,\partial_m P^m\right\}_\Phi = 0$ then $\partial_m P^m$ would be an element of $\mathfrak{N}(2)$. In this event, since $A \in \mathfrak{N}^*(2)$, we would have $T \in \mathfrak{N}(2)$; and the proof used in Case 1 would suffice.

It is probably useful to look at the example considered at the beginning of Case 2 from the viewpoint of the given proof. There we had

$$\Psi \mathcal{O} \Phi - \Phi \mathcal{O}^{+} \Psi = \int_{D^*} \left[ \Psi(x) H^m(x,z) \frac{\partial \Phi(z)}{\partial z^m} + \Phi(x) \Psi(z) \frac{\partial H^m(z,x)}{\partial x^m} \right] dV(z) .$$

After a little manipulation, we obtain

$$\Psi \mathcal{O} \Phi - \Phi \mathcal{O}^{+} \Psi = \frac{\partial}{\partial x^m} \int_{D^*} \Phi(x) \Psi(z) H^m(z,x)\, dV(z)$$

$$+ \int_{D^*} [\Psi(x) H^m(x,z)\, \partial_m \Phi(z) - \Psi(z) H^m(z,x)\, \partial_m \Phi(x)]\, dV(z)$$

so that

$$P^m = \Phi(x) \int_{D^*} \Psi(z) H^m(z,x)\, dV(z)$$

and

$$A = \int_{D^*} [\Psi(x) H^m(x,z)\, \partial_m \Phi(z) - \Psi(z) H^m(z,x)\, \partial_m \Phi(x)]\, dV(z) .$$

The verification that $A \in \mathfrak{N}^*(2)$ is trivial, and $\int_{D^*} A\, dV(x) = 0$.

The following result is now immediate:

**Theorem 2.6.** *Any linear integrodifferential equation*

$$\mathcal{O} \Phi = f , \tag{2.56}$$

*where* $\mathcal{O}$ *is a linear, integrodifferential operator of second order, can be embedded in a variational statement. A lagrangian function which embedds* (2.56) *is given by*

$$L = \Psi \mathcal{O} \Phi - \Psi f + \Phi g , \tag{2.57}$$

*in which case the Euler equations are*

$$\{\mathcal{E} | L\}_{\Psi} = \mathcal{O} \Phi - f = 0 \tag{2.58a}$$

*and*

$$\{\mathcal{E} | L\}_{\Phi} = \mathcal{O}^{+} \Psi + g = 0 . \tag{2.58b}$$

Since the null class of the Euler-Lagrange operators is not vacuous, the lagrangian which embeds a given system of Euler equations is not unique: *if* L *embeds a given system of Euler equations then so does every element of the equivalence class* $\hat{L}$. We could have used

$$2L = \Psi\mathcal{O}\Phi + \Phi\mathcal{O}^{+}\Psi + \partial_m P^m - 2\Psi f + 2\Phi g \tag{2.59}$$

in place of (2.57) since

$$\{\mathcal{E}\,|\,2L\}_{\Psi} = \mathcal{O}\Phi + \mathcal{O}\Phi - 2f\,, \quad \left(\left\{\mathcal{E}\,|\,\Phi\mathcal{O}^{+}\Psi + \partial_m P^m\right\}_{\Psi} = \mathcal{O}\phi\right), \tag{2.60}$$

$$\{\mathcal{E}\,|\,2L\}_{\Phi} = \mathcal{O}^{+}\Psi + \mathcal{O}^{+}\Psi + 2g\,, \quad \left(\left\{\mathcal{E}\,|\,\partial_m P^m\right\}_{\Phi} \equiv 0\right). \tag{2.61}$$

Since we are concerned with linear, second order integrodifferential operators, both (2.57) and (2.59) give lagrangian functions containing second derivatives. Our analysis, however, is based primarily on lagrangian functions which contain derivatives of order less than two. The source of the difficulty lies with the term $a^{km}\partial_k\partial_m\Phi$ which appears in the general integrodifferential operator of the second order and its adjoint $\partial_k\partial_m(a^{km}\Psi)$. In the combination given by (2.59), rather than (2.57), a straightforward calculation shows that

$$\begin{aligned}\Psi a^{km}\partial_k\partial_m\Phi + \Phi\partial_k\partial_m(a^{km}\Psi) &= \partial_k\{a^{km}\partial_m(\Phi\Psi)\} + \Phi\Psi\partial_k\partial_m a^{km} \\ &\quad - 2a^{km}\partial_k\Psi\partial_m\Phi + (\partial_k a^{km})(\Phi\partial_m\Psi - \Psi\partial_m\Phi)\,,\end{aligned} \tag{2.62}$$

and hence the only term involving second derivatives in (2.59) is the divergence

$$\partial_k\{a^{km}\partial_m(\Phi\Psi)\}.$$

This term, however, is an element of the null class of the Euler-Lagrange operators with lagrangian functions containing second derivatives. We thus obtain the following prescription: *set*

$$\mathcal{O}\Phi = a^{km}\partial_k\partial_m\Phi + \tilde{\mathcal{O}}\Phi \tag{2.63}$$

*and let* $\partial_m\tilde{P}^m$ *be the bilinear concomitant formed from* $\tilde{\mathcal{O}}$. *Then*

$$\begin{aligned}2L &= \Psi\tilde{\mathcal{O}}\Phi + \Phi\tilde{\mathcal{O}}^{+}\Psi + \partial_m\tilde{P}^m + \Phi\Psi\partial_k\partial_m a^{km} - 2\Psi f + 2\Phi g \\ &\quad + (\partial_k a^{km})(\Phi\partial_m\Psi - \Psi\partial_m\Phi) - 2a^{km}\partial_k\Psi\partial_m\Phi\end{aligned} \tag{2.64}$$

*gives*

$$\{\mathcal{E}\,|\,L\}_{\Psi} = \mathcal{O}\Phi - f\,, \qquad \{\mathcal{E}\,|\,L\}_{\Phi} = \mathcal{O}^{+}\Psi + g \tag{2.65}$$

*and* L *is independent of second derivatives of* $\Phi$ *and* $\Psi$.

The results given above provide a complete solution to the variational embedding of a single, linear, second-order, integro-differential equation.

We now consider systems of linear, second-order, integro-differential equations. A system of linear, integrodifferential equations

$$\mathcal{O}_{\Gamma\Lambda}\Phi_{\Lambda}(x^k) = f_{\Gamma}(x^k) \tag{2.66}$$

is said to be of *second order* if and only if each of the operators $\mathcal{O}_{\Gamma\Lambda}$ is of second order. A collection of linear operators $\mathcal{O}^{+}_{\Gamma\Lambda}$ is said to be the *adjoint* of a collection of linear operators $\mathcal{O}_{\Gamma\Lambda}$ of second order if and only if

$$\begin{aligned}\int_{D^*} &\{\Psi_{\Gamma}\mathcal{O}_{\Gamma\Lambda}\Phi_{\Lambda} - \Phi_{\Gamma}\mathcal{O}^{+}_{\Gamma\Lambda}\Psi_{\Lambda}\}dV(x) \\ &= \int_{D^*} \partial_m P^m(x^k, \Phi_{\Lambda}, \Psi_{\Lambda}, \partial_m\Phi_{\Lambda}, \partial_m\Psi_{\Lambda})\, dV(x)\end{aligned} \tag{2.67}$$

holds for all functions $\Phi_{\Lambda}$ and $\Psi_{\Lambda}$ in some specified classes $R_1$ and $R_2$ of $N$-tuples of functions on $D^*$ and the $P$'s are multilinear in $\Phi_{\Lambda}, \Psi_{\Lambda}, \partial_m\Phi_{\Lambda}, \partial_m\Psi_{\Lambda}$.

**Theorem 2.7.** *Let* $\mathcal{O}_{\Gamma\Lambda}$ *be a given system of linear, second order, integrodifferential operators. Then we have*

$$\{\mathcal{E}\,|\Psi_{\Gamma}\mathcal{O}_{\Gamma\Lambda}\Phi_{\Lambda}\}_{\Phi_{\Sigma}} = \mathcal{O}^{+}_{\Sigma\Lambda}\Psi_{\Lambda} \tag{2.68a}$$

$$\{\mathcal{E}\,|\Psi_{\Gamma}\mathcal{O}_{\Gamma\Lambda}\Phi_{\Lambda}\}_{\Psi_{\Sigma}} = \mathcal{O}_{\Sigma\Lambda}\Phi_{\Lambda} \tag{2.68b}$$

and

$$\{\mathcal{E}\,|\,\Phi_{\Gamma}\mathcal{O}^{+}_{\Gamma\Lambda}\Psi_{\Lambda} + \partial_m P^m\}_{\Phi_{\Sigma}} = \mathcal{O}^{+}_{\Sigma\Lambda}\Psi_{\Lambda} \tag{2.69a}$$

$$\{\mathcal{E}\,|\,\Phi_{\Gamma}\mathcal{O}^{+}_{\Gamma\Lambda}\Psi_{\Lambda} + \partial_m P^m\}_{\Psi_{\Sigma}} = \mathcal{O}_{\Sigma\Lambda}\Phi_{\Lambda}\,. \tag{2.69b}$$

**Proof.** The proof follows the same lines as that given for a single linear, second-order, integrodifferential operator and is left to the reader. Note that, if there are only differential operators present, then $\partial_m P^m$ is annihilated by both $\{\mathcal{E}\,|\;\;\}_{\Psi_{\Sigma}}$ and $\{\mathcal{E}\,|\;\;\}_{\Phi_{\Sigma}}$.

The following result is now obvious.

**Theorem 2.8.** *Any system of linear, second-order, integrodifferential equations*

$$\mathcal{O}_{\Gamma\Lambda}\Phi_{\Lambda}(x^m) = f_{\Gamma}(x^m) \tag{2.70}$$

*can be embedded in a variational statement. A lagrangian function that accomplishes this embedding is given by*

$$L = \Psi_{\Gamma}\mathcal{O}_{\Gamma\Lambda}\Phi_{\Lambda} - \Psi_{\Gamma}f_{\Gamma} + \Phi_{\Gamma}g_{\Gamma} \tag{2.71}$$

*or by*

$$2L = \Psi_{\Gamma}\mathcal{O}_{\Gamma\Lambda}\Phi_{\Lambda} + \Phi_{\Gamma}\mathcal{O}^{+}_{\Gamma\Lambda}\Psi_{\Lambda} - 2\Psi_{\Gamma}f_{\Gamma} + 2\Phi_{\Gamma}g_{\Gamma} + \partial_m P^m. \tag{2.72}$$

The student is strongly advised to make up several problems with $N = 2$ and to construct $L$, $\mathcal{O}^{+}_{\Gamma\Lambda}$, and the bilinear concomitant for the assumed forms of $\mathcal{O}_{\Gamma\Lambda}$.

Another situation arises in a natural fashion in the study of differential geometry and continuous groups, namely, that in which we are given a system of linear, second-order, integrodifferential equations involving only one function. Such systems are represented by

$$\mathcal{O}_r\Phi(x^k) = f_r(x^k) \tag{2.73}$$

where the index $r$ usually rangaes from one through a number less than or equal to the number of independent variables and $\mathcal{O}_r$ is a linear, second-order, integrodifferential operator for each value of $r$. The adjoints of the operators $\mathcal{O}_r$ are defined as follows:

$$\int_{D^*}\{\Psi^r\mathcal{O}_r\Phi - \Phi\mathcal{O}^{+}_r\Psi^r\}dV(x) = \int_{D^*}\partial_m P^m\, dV(x)\,. \tag{2.74}$$

**Theorem 2.9.** *Any system of second-order, integrodifferential equations in one unknown can be embedded in a variational statement. If the system is given by* (2.73), *then a lagrangian function that accomplishes the embedding is given by*

$$L = \Psi^r\mathcal{O}_r\Phi - \Psi^r f_r. \tag{2.75}$$

*Further, we have*

$$\left\{\mathcal{E} \mid \Psi^r \mathcal{O}_r \Phi\right\}_\Phi = \mathcal{O}_r^+ \Psi^r , \tag{2.76a}$$

$$\left\{\mathcal{E} \mid \Psi^r \mathcal{O}_r \Phi\right\}_{\Psi^r} = \mathcal{O}_r \Phi , \tag{2.76b}$$

$$\left\{\mathcal{E} \mid \Phi \mathcal{O}_r^+ \Psi^r + \partial_m P^m\right\}_{\Psi^s} = \mathcal{O}_s \Phi , \tag{2.77a}$$

$$\left\{\mathcal{E} \mid \Phi \mathcal{O}_r^+ \Psi^r + \partial_m P^m\right\}_\Phi = \mathcal{O}_r^+ \Psi^r , \tag{2.77b}$$

*and*

$$\{\mathcal{E} \mid L\}_{\Psi^r} = \mathcal{O}_r \Phi - f_r , \qquad \{\mathcal{E} \mid L\}_\Phi = \mathcal{O}_r^+ \Psi^r . \tag{2.78}$$

The following result is an immediate consequence of combining the above statement with the definition

$$\int_{D^*} \{\Psi^r \mathcal{O}_{r\Lambda\Gamma} \Phi_\Gamma - \Phi_\Gamma \mathcal{O}^+_{r\Gamma\Lambda} \Psi^r_\Lambda\} dV(x) = \int_{D^*} \partial_m P^m \, dV(x) .$$

**Theorem 2.10.** *Any system of* pN *linear, second-order, integro-differential equations in* N *unknowns*

$$\mathcal{O}_{r\Lambda\Gamma} \Phi_\Gamma(x^k) = f_{r\Lambda}(x^k) \qquad r = 1, \ldots, p , \tag{2.79}$$

*can be embedded in a variational statement. A lagrangian function that accomplishes the embedding is given by*

$$L = \Psi^r_\Lambda (\mathcal{O}_{r\Lambda\Gamma} \Phi_\Gamma - f_{r\Lambda}) . \tag{2.80}$$

*Thus, we have*

$$\left\{\mathcal{E} \mid \Psi^r_\Lambda \mathcal{O}_{r\Lambda\Gamma} \Phi_\Gamma\right\}_{\Phi_\Sigma} = \mathcal{O}^+_{r\Sigma\Lambda} \Psi^r_\Lambda , \tag{2.81a}$$

$$\left\{\mathcal{E} \mid \Psi^r_\Lambda \mathcal{O}_{r\Lambda\Gamma} \Phi_\Gamma\right\}_{\Psi^s_\Sigma} = \mathcal{O}_{s\Sigma\Lambda} \Phi_\Lambda , \tag{2.81b}$$

$$\left\{\mathcal{E} \mid \Phi_\Lambda \mathcal{O}^+_{r\Lambda\Gamma} \Psi^r_\Gamma + \partial_m P^m\right\}_{\Psi^s_\Sigma} = \mathcal{O}_{s\Sigma\Lambda} \Phi_\Lambda , \tag{2.82a}$$

$$\left\{\mathcal{E} \mid \Phi_\Lambda \mathcal{O}^+_{r\Lambda\Gamma} \Psi^r_\Gamma + \partial_m P^m\right\}_{\Phi_\Sigma} = \mathcal{O}^+_{r\Sigma\Lambda} \Psi^r_\Lambda , \tag{2.82b}$$

*so that*

$$\{\mathcal{E} \mid L\}_{\Phi_\Gamma} = \mathcal{O}^+_{r\Gamma\Lambda} \Psi^r_\Lambda , \qquad \{\mathcal{E} \mid L\}_{\Psi^r_\Lambda} = \mathcal{O}_{r\Lambda\Gamma} \Phi_\Gamma - f_{r\Lambda} . \tag{2.83}$$

Note: The lagrangian that accomplishes an embedding of a given system of equations is *not unique*; if $L$ is such a lagrangian, then $L + A$ also embeds the given equations for all $A \in \mathfrak{N}(N)$.

In contrast to the results obtained in Chap. 1, we have been able to embed any collection of second-order integrodifferential equations in a variational statement. The embedding comes at an apparent high price, however, for we require the "adjoint functions" as well as the original functions involved in the given system of integrodifferential equations. In actuality, the price of additional unknowns is worthwhile, for it allows us to obtain necessary conditions for the existence of solutions to the given system of integrodifferential equations. This comes about as follows. In the general case, we start with the linear, second-order system of integrodifferential equations

$$\mathcal{O}_{r\Gamma\Lambda}\Phi_\Lambda(x^m) = f_{r\Gamma} \tag{2.84}$$

and obtain the lagrangian function

$$L = \Psi^r_\Gamma(\mathcal{O}_{r\Gamma\Lambda}\Phi_\Lambda - f_{r\Gamma}) . \tag{2.85}$$

This gives us the functional

$$J[\Phi_\Lambda, \Psi^r_\Gamma] = \int_{D^*} \{\Psi^r_\Gamma(\mathcal{O}_{r\Gamma\Lambda}\Phi_\Lambda - f_{r\Gamma})\} dV(x) \tag{2.86}$$

and

$$\frac{\delta J}{\delta \Psi_\Gamma{}^r} = 0 \Rightarrow \mathcal{O}_{r\Gamma\Lambda}\Phi_\Lambda(x^m) = f_{r\Gamma} \tag{2.87}$$

$$\frac{\delta J}{\delta \Phi_\Gamma} = 0 \Rightarrow \mathcal{O}^+_{r\Gamma\Lambda}\Psi^r_\Lambda = 0 . \tag{2.88}$$

On the other hand, (2.86) and (2.84) show that

$$J[\Phi_\Lambda, \Psi^r_\Gamma] = 0 \tag{2.89}$$

holds for any $\{\Phi_\Lambda(x^m)\}$ which are solutions of the system (2.84) and for all $\{\Psi^r_\Lambda\}$. Hence, not only is the functional $J[\Phi_\Lambda, \Psi^r_\Gamma]$ stationary with respect to $\{\Psi^r_\Lambda\}$ for any solution of (2.84), it has the value zero for any solution of (2.84) and for all $\{\Psi^r_\Lambda\}$.

As an example of the use of this observation, consider the problem

$$\partial_k \Phi(x^m) = f_k(x^m) , \qquad \Phi \in C^2 , \; f_k \in C^1 , \tag{2.90}$$

so that

$$\mathcal{O}_k \equiv \partial_k \quad , \quad \mathcal{O}_k^+ \equiv -\partial_k . \tag{2.91}$$

We then have

$$J[\Phi, \Psi^k] = \int_{D^*} \Psi^k (\partial_k \Phi - f_k) dV(x) , \tag{2.92}$$

and the Euler equations for (2.92) are

$$\{\mathcal{E} | L\}_{\Psi^k} = \partial_k \Phi - f_k = 0 \tag{2.93}$$

$$\{\mathcal{E} | L\}_{\Phi} = -\partial_k \Psi^k = 0 . \tag{2.94}$$

Hence, if $J[\Phi, \Psi^k]$ is stationary with respect to the choice of $\{\Psi_k\}$ (i.e., if a solution of (2.90) exists), then we have

$$J[\Phi, \Psi^k] = 0 \tag{2.95}$$

for all functions $\{\Psi^k\}$.

Let us now choose $\Phi$ so as to stationarize $J[\Phi, \Psi^k]$ with respect to $\Phi$. By (2.94), this means that $\{\Psi^k\}$ must satisfy

$$\partial_k \Psi^k = 0 . \tag{2.96}$$

A general solution of this equation is given by

$$\Psi^k = \partial_m S^{mk} \tag{2.97}$$

where $\{S^{mk}\}$ is an arbitrary collection of $n^2$ functions of class $C^2$ such that

$$S^{mk} = -S^{km} . \tag{2.98}$$

If we evaluate $J[\Phi, \Psi^k]$ for this $\{\Psi^k\}$, we have

$$J[\Phi, \Psi^k] = \int_{D^*} (\partial_m S^{mk})(\partial_k \Phi - f_k) dV(x)$$

$$= \int_{D^*} \partial_m \{S^{mk}(\partial_k \Phi - f_k)\} dV(x) - \int_{D^*} S^{mk} \{\partial_m f_k - \partial_m \partial_k \Phi\} dV(x) .$$

Since $S^{mk} \in C^2$ and (2.98) holds, we accordingly have

$$J[\Phi, \Psi^k] = -\frac{1}{2}\int_{D^*} S^{mk}(\partial_m f_k - \partial_k f_m)\, dV(x) + \int_{D^*} \partial_m \{S^{mk}(\partial_k \Phi - f_k)\}\, dV(x) . \tag{2.99}$$

We saw, however, that $J[\Phi, \Psi^k] = 0$ for all $\{\Psi^k\}$ whenever $\Phi$ satisfied (2.90). In this case, (2.99) gives

$$J[\Phi, \Psi^k] = -\int_{D^*} \frac{1}{2} S^{mk}(\partial_m f_k - \partial_k f_m)\, dV(x) \tag{2.100}$$

for all $S^{mk}$ of class $C^2$ such that (2.98) holds. We thus obtain a contradiction to the assumption that $\Phi$ satisfies (2.90) (i.e., a solution of (2.90) exists) unless

$$\partial_m f_k = \partial_k f_m \tag{2.101}$$

holds. Hence (2.101) are necessary conditions for the existence of solutions to (2.90). The conditions (2.101) are, however, the classical integrability conditions for the system (2.90). In this problem (2.101) could have been obtained directly from (2.90), but in more complicated situations, the above method works -- even in those cases in which there are integral operations.

Consider the problem in which we wish to solve

$$\int_{D^*} K(x^m, z^m)\, \Phi(z^m)\, dV(z) = f(x) . \tag{2.102}$$

The lagrangian is

$$L = \Psi(x^m) \int_{D^*} K(x^m, z^m)\, \Phi(z^m)\, dV(z^m) - \Psi(x^m)\, f(x^m) \tag{2.103}$$

and, hence

$$J[\Phi, \Psi] = -\int_{D^*} \Psi(x^m)\, f(x^m)\, dV(x) + \iint_{D^*} \Psi(x^m)\, K(x^m, z^m)\, \Phi(z^m)\, dV(z) dV(x) \tag{2.104}$$

has the value of zero for all functions $\Psi$ whenever $\Phi$ satisfies (2.102). If we choose $\Phi$ so as to stationarize $J[\Phi, \Psi]$, we have

$$\int_{D^*} \Psi(z^m)\, K(z^m, x^m)\, dV(z) = \{\mathcal{E} \mid L\}_{\Phi} = 0 . \tag{2.105}$$

An evaluation of $J[\Phi, \Psi]$ for such $\Psi$ gives

$$J[\Phi, \Psi] = -\int \Psi f dV(x) . \tag{2.106}$$

We thus have a contradiction unless

$$\int_{D^*} \Psi(x^m) f(x^m)\, dV(x) = 0 \tag{2.107}$$

*holds for all* $\Psi(x^m)$ *such that*

$$\int_{D^*} \Psi(z^m) K(z^m, x^m)\, dV(x) = 0 . \tag{2.108}$$

As a final example, consider the problem of an integrodifferential equation analogous to (2.90):

$$\int_{D^*} H_s{}^k(x^m, z^m) \frac{\partial \Phi(z^m)}{\partial z^k}\, dV(z) = f_s(x^m) , \qquad s = 1, \ldots, n . \tag{2.109}$$

We then have

$$\mathcal{O}_s \Phi = \int_{D^*} H_s{}^k(x^m, z^m)\, \partial_k \Phi(z^m)\, dV(z) , \tag{2.110}$$

$$\mathcal{O}_s^+ \Psi^s = -\frac{\partial}{\partial x^k} \int_{D^*} \Psi^s(z^m) H_s{}^k(z^m, x^m)\, dV(z) , \tag{2.111}$$

with the bilinear concomitant

$$p^k = \int_{D^*} \Psi^s(z^m) H_s{}^k(z^m, x^m)\, \Phi(x^m)\, dV(z) , \tag{2.112}$$

The functional which embeds this problem in a variational statement is given by

$$J[\Phi, \Psi^s] = -\int_{D^*} \Psi^s f_s\, dV + \iint_{D^*} \Psi^s(x^m) H_s{}^k(x^m, z^m)\, \partial_k \Phi(z^m)\, dV(z)\, dV(x)$$

$$= -\int_{D^*} \Psi^s f_s\, dV + \iint_{D^*} \Psi^s(z^m) H_s{}^k(z^m, x^m)\, \partial_k \Phi(x^m)\, dV(x)\, dV(z)$$

$$= -\int_{D^*} \Psi^s f_s \, dV + \int_{D^*} \left[ \frac{\partial}{\partial x^k} \int_{D^*} \Psi^s(z^m) H_s{}^k(z^m, x^m) \cdot \Phi(x^m) dV(z) \right] dV(x)$$

$$- \int_{D^*} \Phi(x) \frac{\partial}{\partial x^k} \left[ \int_{D^*} \Psi^s(z) H_s{}^k(z^m, x^m) \, dV(z) \right] dV(x)$$

$$= -\int_{D^*} \Psi^s f_s \, dV(x) + \int_{D^*} \frac{\partial p^k}{\partial x^k} dV(x) + \int_{D^*} \Phi(x) \mathcal{O}_s^+ \Psi^s \, dV(x) . \tag{2.113}$$

For any $\Phi$ which satisfies (2.109) and all $\Psi^s$, we have $J[\Phi, \Psi^s] = 0$.

On the other hand, if we choose $\Phi$ so that $J[\Phi, \Psi^s]$ is stationary with respect to $\Phi$, then

$$\{\mathcal{E} | L\}_\Phi = \mathcal{O}_s^+ \Psi^s = 0 , \tag{2.114}$$

and hence, by (2.111), $\Psi^s$ must satisfy

$$\int_{D^*} \Psi^s(z^m) H_s{}^k(z^m, x^m) \, dV(z) = \partial_r S^{rk}(x^m) \tag{2.115}$$

for $\{S^{mk}(x)\} \in C^2$ and $S^{mk} = -S^{km}$. The system (2.115) is almost identical to the previous example; in fact, if we define $\mathcal{O}_s{}^k$ by

$$\mathcal{O}_s{}^k \Psi^s(x^m) = \int_{D^*} \Psi^s(z^m) H_s{}^k(z^m, x^m) \, dV(z) ,$$

we have

$$\mathcal{O}_s^{+k} \eta_k(x^m) = \int_{D^*} H_s{}^k(x^m, z^m) \, \eta_k(z^m) \, dV(z) ,$$

since

$$\tilde{L} = \eta_k(x^m) \mathcal{O}_s{}^k \Psi^s \Rightarrow \{\mathcal{E} | \tilde{L}\}_{\Psi^s} = \mathcal{O}_s^{+k} \eta_k(x^m) .$$

By the same reasoning as in the previous example, a solution of (2.115) exists only if

$$\int_{D^*} [\partial_r S^{rk}(x^m)] \, \eta_k(x^m) \, dV(x) = 0 \tag{2.116}$$

for all $\eta_k(s)$ which satisfy

$$\mathcal{O}_s^{+k}\eta_k = \int_{D^*} H_s{}^k(x^m, z^m)\,\eta_k(z^m)\,dV(z) = 0. \tag{2.117}$$

Returning to the present problem, if we multiply (2.115) by $\Phi(x^m)$ and if (2.112) is used, we have

$$\Phi(x^m)\,\partial_m S^{mk}(x^m) = p^k. \tag{2.118}$$

If $J[\Phi, \Psi^s]$ is evaluated for $\Psi^k$ which satisfies (2.115), we have

$$J[\Phi, \Psi^s] = -\int_{D^*} \Psi^s f_s\,dV(x) + \int_{\partial D} \Phi\partial_m S^{mk}\,dS_k(x). \tag{2.119}$$

However, when $\Phi(x^m)$ satisfies (2.109), we have $J[\Phi, \Psi^s] = 0$ for all $\Psi^s$, and hence there is a contradiction unless the right-hand side of (2.119) vanishes. Collecting the various results, we have the following conclusion. *A solution of*

$$\int_{D^*} H_s{}^k(x^m, z^m)\,\frac{\partial\Phi(z^m)}{\partial z^k}\,dV(z) = f_s(x^m)$$

*exists only if*

$$\int_{D^*} \Psi^s f_s\,dV(x) = \int_{\partial D} p^k\,dS_k(x) = \int_{\partial D} \Phi(x)\,\partial_m S^{mk}(x)\,dS_k(x) \tag{2.120}$$

*for all* $\Psi^s$ *such that*

$$\int_{D^*} \Psi^s(z^m)\,H_s{}^k(z^m, x^m)\,dV(z) = \partial_m S^{mk}(x) \tag{2.121}$$

*and for all* $S^{mk} \in C^2$ *such that* (1) $S^{mk} = -S^{km}$, *and* (2) *that*

$$\int_{D^*} (\partial_m S^{mk})\,\eta_k\,dV(x) = 0 \tag{2.122}$$

*for all* $\eta_k(x^m)$ *such that*

$$\int_{D^*} H_s{}^k(x^m, z^m)\,\eta_k(z^m)\,dV(z) = 0. \tag{2.123}$$

Hopefully, the considerations of this section show the unity that can be given to the study of integrodifferential equations through the use of variational considerations. First, it is seen that the Euler-Lagrange operator is an automatic adjoint calculator which works whether differential operators, integral operators, or integrodifferential operators in all of their various combinations are being considered. Second, the variational embedding of a given linear system of operator equations leads to necessary conditions for existence of solutions. These necessary conditions take the form of requirements between the inhomogeneous part of the equations and solutions of the homogeneous adjoint equations. A unified and direct method of obtaining necessary conditions for existence is thus obtained for overdetermined systems of differential equations, integral equations and integrodifferential equations. An important aspect which emerges from these considerations, is that the necessary conditions for overdetermined differential equations, integral equations, and integrodifferential equations branch from a single basic conception. This is in contrast to usual methods which either point out dissimilarities or indicate totally different occurrences in each of the cases.

The student is advised to examine a number of cases in addition to those given above, in order to gain understanding of integrability or existence conditions and insight into new and interesting problems. Sufficiency or unicity conditions are not claimed, but, within the context of physical problems and theories, existence conditions (necessity rather than sufficiency) more often than not ensure understanding and success.

## 2.5. MOMENTUM-ENERGY COMPLEXES

A set of fundamental integrodifferential identities is obtained in this section. These identities involve an ordered array of functions and functionals which, because of the physical origin of the considerations leading to their historical development, is referred to as a momentum-energy complex.

We start by recalling and augmenting the notation introduced in Sec. 2.2. The functions

$$g_a = g_a[x^m, z^m, \Phi_\Lambda(z^m), \partial_k \Phi_\Lambda(z^m)] , \qquad a = 1, \ldots, q, \qquad (2.124)$$

are assumed to be of class $C^2$ in their indicated arguments, and the functions $g_a^*$ are defined by

$$g_a^* = g_a[z^m, x^m, \Phi_\Lambda(x^m), \partial_k \Phi_\Lambda(x^m)] , \qquad a = 1, \ldots, q. \qquad (2.125)$$

These functions serve to define two functional sets. The first,

$$k_a(x^m, \{\Phi_\Lambda\}) = \int_{D^*} g_a[x^m, z^m, \Phi_\Lambda(z^m), \partial_k \Phi_\Lambda(z^m)]\, dV(z), \tag{2.126}$$

treats $x$ as parameters, and the second

$$k_a^*(z^m, \{\Phi_\Lambda\}) = \int_{D^*} g_a^*[z^m, x^m, \Phi_\Lambda(x^m), \partial_k \Phi_\Lambda(x^m)]\, dV(x) \tag{2.127}$$

treats $z$ as parameters. If we differentiate $g_a^*$ with respect to $x$, we have

$$\frac{\partial g_a^*}{\partial x^k} = \partial_k^* g_a^* + \frac{\partial g_a^*}{\partial \Phi_\Lambda} \partial_k \Phi_\Lambda + \frac{\partial g_a^*}{\partial(\partial_m \Phi_\Lambda)} \partial_k \partial_m \Phi_\Lambda \tag{2.128}$$

where $\partial_k^* g_a^*$ denotes the derivative of $g_a^*$ with respect to $x^k$ for constant $\Phi_\Lambda(x^m)$, $\partial_k \Phi_\Lambda(x^m)$, $z^m$. With this notation, we establish the following preliminary result.

**Lemma 2.2.** *Define the $n^2 q$ quantities*

$$w_{ak}{}^m = w_{ak}{}^m[z^t, x^t, \Phi_\Lambda(x^t), \partial_s \Phi_\Lambda(x^t); g_a^*]$$

*by*

$$w_{ak}{}^m = \frac{\partial g_a^*}{\partial(\partial_m \Phi_\Lambda)} \partial_k \Phi_\Lambda - \delta_k{}^m g_a^*, \qquad \delta_k{}^m = \begin{cases} 1 & \text{if } k = m \\ 0 & \text{if } k \neq 0 . \end{cases} \tag{2.129}$$

*The relations*

$$\frac{\partial w_{ak}{}^m}{\partial x^m} = -\{e \mid g_a^*\}_{\Phi_\Lambda} \partial_k \Phi_\Lambda(x) - \partial_k^* g_a^* \tag{2.130}$$

*are identically satisfied on $\mathfrak{D}_1(D^*, N)$, and the relations*

$$\frac{\partial w_{ak}^m}{\partial x^m} = -\partial_k^* g_a^* \tag{2.131}$$

*are identically satisfied on the subset of* $\mathfrak{D}_1(D^*, N)$ *for which*

$$\left\{e \mid g_a^*\right\}_{\Phi_\Lambda} = 0 \tag{2.132}$$

*holds.*

**Proof.** The proof is by direct calculation:

$$\frac{\partial w_{ak}{}^m}{\partial x^m} = \frac{\partial g_a^*}{\partial(\partial_m \Phi_\Lambda)} \partial_m \partial_k \Phi_\Lambda + \frac{\partial}{\partial x^m}\left[\frac{\partial g_a^*}{\partial(\partial_m \Phi_\Lambda)}\right] \partial_k \Phi_\Lambda - \frac{\partial g_a^*}{\partial x^k}$$

$$= \frac{\partial g_a^*}{\partial(\partial_m \Phi_\Lambda)} \partial_m \partial_k \Phi_\Lambda + \frac{\partial}{\partial x^m}\left[\frac{\partial g_a^*}{\partial(\partial_m \Phi_\Lambda)}\right] \partial_k \Phi_\Lambda - \partial_k^* g_a^*$$

$$- \frac{\partial g_a^*}{\partial \Phi_\Lambda} \partial_k \Phi_\Lambda - \frac{\partial g_a^*}{\partial(\partial_m \Phi_\Lambda)} \partial_k \partial_m \Phi_\Lambda$$

$$= \left\{\frac{\partial}{\partial x^m}\left[\frac{\partial g_a^*}{\partial(\partial_m \Phi_\Lambda)}\right] - \frac{\partial g_a^*}{\partial \Phi_\Lambda}\right\} \partial_k \Phi_\Lambda - \partial_k^* g_a^* = -\left\{e \mid g_a^*\right\}_{\Phi_\Lambda} \partial_k \Phi_\Lambda - \partial_k^* g_a^* .$$

As before, let $L = L[x^m, \Phi_\Lambda(x^m), \partial_k \Phi_\Lambda(x^m), k_a(x^m, \{\Phi_\Lambda\})]$, and introduce the notation

$$\frac{\partial L}{\partial x^k} = \partial_k^* L + \frac{\partial L}{\partial \Phi_\Lambda} \partial_k \Phi_\Lambda + \frac{\partial L}{\partial(\partial_m \Phi_\Lambda)} \partial_k \partial_m \Phi_\Lambda + \frac{\partial L}{\partial k_a} \frac{\partial k_a}{\partial x^m}, \tag{2.133}$$

in which $\partial_k^* L$ denotes the derivative of $L$ with respect to $x^k$ for constant $\Phi_\Lambda, \partial_m \Phi_\Lambda, k_a$ and $(\partial L/\partial k_a)(z^m)$ denotes the derivative of $L$ with respect to $k_a$ and then replace $x^m$ by $z^m$ throughout.

Adopting present usage, we say that a relation involving an equality is a *strong identity* on $\mathfrak{D}_1(D^*, N)$ if the relation holds for all elements of $\mathfrak{D}_1(D^*, N)$ and that a relation involving an equality is a *weak identity* with respect to $L$ if the relation holds for all elements of $S(L)$. We can now state the basic theorem concerning momentum-energy complexes.

**Theorem 2.11.** *For a given function* L *define the* $n^2$ *functions* $W_k{}^m$ *by the relations*

$$W_k{}^m = \frac{\partial L}{\partial(\partial_m \Phi_\Lambda)} \partial_k \Phi_\Lambda + \int_{D^*} w_{ak}{}^m \frac{\partial L}{\partial k_a}(z^t)\, dV(z) - \delta_k{}^m L . \quad (2.134)$$

*This collection of functions constitutes the momentum-energy complex based on L. This momentum-energy complex gives rise to the strong identities*

$$\frac{\partial W_k{}^m}{\partial x^m} = -\{\mathcal{E}\,|\,L\}_{\Phi_\Lambda} \partial_k \Phi_\Lambda - \partial_k^* L - \int_{D^*} \left[ \frac{\partial L}{\partial k_a}(z^t)\, \partial_k^* g_a^* + \frac{\partial L}{\partial k_a} \frac{\partial g_a}{\partial x^k} \right] dV(z) \quad (2.135)$$

*and to the weak identities*

$$\frac{\partial W_k{}^m}{\partial x^m} = -\partial_k^* L - \int_{D^*} \left[ \frac{\partial L}{\partial k_a}(z^t)\, \partial_k^* g_a^* + \frac{\partial L}{\partial k_a} \frac{\partial g_a}{\partial x^k} \right] dV(z) . \quad (2.136)$$

**Proof.** The proof is by direct calculation. Since $(\partial L/\partial k_a)(z^t)$ is independent of $x^m$, we have

$$\frac{\partial W_k{}^m}{\partial x^m} = \frac{\partial L}{\partial(\partial_m \Phi_\Lambda)} \partial_m \partial_k \Phi_\Lambda + \partial_k \Phi_\Lambda \frac{\partial}{\partial x^m} \frac{\partial L}{\partial(\partial_m \Phi_\Lambda)} + \int_{D^*} \frac{\partial w_{ak}{}^m}{\partial x^m} \frac{\partial L}{\partial k_a}(z^t)\, dV(z)$$
$$- \partial_k^* L - \frac{\partial L}{\partial \Phi_\Lambda} \partial_k \Phi_\Lambda - \frac{\partial L}{\partial(\partial_m \Phi_\Lambda)} \partial_k \partial_m \Phi_\Lambda - \frac{\partial L}{\partial k_a} \frac{\partial k_a}{\partial x^m} . \quad (2.137)$$

When (2.130) of the Lemma 2.2 is used and (2.126) is used to obtain

$$\frac{\partial L}{\partial k_a} \frac{\partial k_a}{\partial x^m} = \frac{\partial L}{\partial k_a} \int_{D^*} \frac{\partial g_a}{\partial x^m} dV(z) = \int_{D^*} \frac{\partial L}{\partial k_a} \frac{\partial g_a}{\partial x^m} dV(z) ,$$

since $\partial L/\partial k_a$ is a function of $x^m$ only, then (2.137) becomes

$$\frac{\partial W_k{}^m}{\partial x^m} = \partial_k \Phi_\Lambda \left\{ \frac{\partial}{\partial x^m} \left[ \frac{\partial L}{\partial(\partial_m \Phi_\Lambda)} \right] - \frac{\partial L}{\partial \Phi_\Lambda} \right\} - \partial_k^* L$$
$$- \int_{D^*} \left[ \frac{\partial L}{\partial k_a} \frac{\partial g_a}{\partial x^m} + \{e\,|\,g_a^*\}_{\Phi_\Lambda} \partial_k \Phi_\Lambda(x) \frac{\partial L}{\partial k_a}(z^t) + \frac{\partial L}{\partial k_a}(z^t)\, \partial_k^* g_a^* \right] dV(z) .$$

Now, $\partial_k \Phi_\Lambda(x)$ is independent of $(z^m)$ and hence can be taken outside of the $z$-integral. When this is done and (2.13) is used, we obtain (2.135) for any $\Phi_\Lambda \in \mathfrak{D}_1(D^*, N)$. Equation (2.136) then follows from (2.135) for any $\Phi_\Lambda \in S(L)$ since $\{\mathcal{E} \mid L\}_{\Phi_\Lambda} = 0$ for all elements of $S(L)$.

The following examples will help to clarify what is contained (and hidden) within the notation we have introduced. First, consider the situation in which $L$ is a sum of terms, one of which is

$$\Phi(x^m)\, k(x^m; \Phi) \;, \qquad k(x^m; \Phi) = \int_{D^*} K(x^m, z^m)\, \Phi(z^m)\, dV(z) \;,$$

and $N = q = 1$. We then have

$$g = K(x^m, z^m)\, \Phi(z^m) \;, \qquad \frac{\partial g}{\partial x^k} = \frac{\partial K(x^m, z^m)}{\partial x^k}\, \Phi(z^m) \;,$$

$$g^* = K(z^m, x^m)\, \Phi(x^m) \;, \qquad \partial^*_k g^* = \frac{\partial K(z^m, x^m)}{\partial x^k}\, \Phi(x^m)$$

$$\frac{\partial L}{\partial k} = \Phi(x^m) \;, \qquad \frac{\partial L}{\partial k}(z^m) = \Phi(z^m) \;,$$

and consequently

$$w_k{}^m = -\delta_k{}^m g^* = -\delta_k{}^m K(z^m, x^m)\, \Phi(x^m)$$

$$W_k{}^m = \int_{D^*} w_m{}^k \frac{\partial L}{\partial k}(z^m)\, dV(z) - \delta_k{}^m L$$

$$= -\delta_k{}^m \Phi(x^m) \int_{D^*} [K(z^m, x^m) + K(x^m, z^m)]\, \Phi(z^m)\, dV(z) \,.$$

On the other hand, (2.136) gives, on $S(L)$,

$$\frac{\partial W_k{}^m}{\partial x^m} = -\int_{D^*} \left[ \frac{\partial L}{\partial k}(z)\, \partial^*_k g^* + \frac{\partial L}{\partial k} \frac{\partial g}{\partial x^k} \right] dV(z)$$

$$= \Phi(x^m) \int_{D^*} \left[ \frac{\partial K(z^m, x^m)}{\partial x^k} + \frac{\partial K(x^m, z^m)}{\partial x^k} \right] \Phi(z^m)\, dV(z) \,.$$

Hence, if $K(z^m, x^m) = -K(x^m, z^m)$ then $W_k{}^m \equiv 0$, $\partial W_k{}^m / \partial x^m \equiv 0$ and if $K(x^m, z^m) = K(z^m, x^m)$ then $W_k{}^m = 2\delta_k{}^m \Phi(x^m) \int K(x^m, z^m)\, \Phi(z)\, dV(z)$ $= -2\delta_k{}^m L$ and

$$\frac{\partial W_k{}^m}{\partial x^m} = -2\Phi(x^m)\int_{D^*} \frac{\partial K(x^m, z^m)}{\partial x^k} \Phi(z^m)\, dV(z) \quad \text{for} \quad \forall \Phi(x^m) \in S(L).$$

As a second example, we take

$$L = \frac{1}{2}\Phi^2 + \Phi\int_{D^*} K^t(x^m, z^m)\, \partial_t \Phi(z_m)\, dV(z),$$

that is, $N = q = 1$. We accordingly have

$$g = K^t(x^m, z^m)\, \partial_t \Phi(z^m) \quad , \quad \frac{\partial g}{\partial x^k} = \frac{\partial K^t(x^m, z^m)}{\partial x^k} \partial_t \Phi(z^m),$$

$$g^* = K^t(z^m, x^m)\, \partial_t \Phi(x^m) \quad , \quad \partial_k^* g^* = \frac{\partial K^t(z^m, x^m)}{\partial x^m} \partial_t \Phi(x^m),$$

$$\frac{\partial L}{\partial k} = \Phi(x^m) \quad , \quad \frac{\partial L}{\partial k}(z^m) = \Phi(z^m),$$

$$\frac{\partial g^*}{\partial(\partial_m \Phi)} = K^m(z^k, x^k).$$

Substituting these into the previous formulas gives

$$w_k{}^m = \partial_k \Phi(x^m)\, K^m(z^m, x^m) - \delta_k{}^m K^t(z^m, x^m)\, \partial_t \Phi(x^m)$$

$$W_k{}^m = -\delta_k{}^m \left[\frac{1}{2}\Phi^2 + \Phi\int_{D^*} K^t(x^m, z^m)\, \partial_t \Phi(z^m)\, dV(z)\right]$$
$$+ \int_D [\partial_k \Phi(x^m)\, K^m(z^m, x^m) - \delta_k{}^m K^t(z^m, x^m)]\, \Phi(x^m)\, dV(z).$$

Thus, on $S(L)$, we have

$$\frac{\partial W_k{}^m}{\partial x^m} = -\int_{D^*}\left[\Phi(z^m)\frac{\partial K^t(z^m, x^m)}{\partial x^k}\partial_t \Phi(x^m) + \Phi(x^m)\frac{\partial K^t(x^m, z^m)}{\partial x^k}\partial_t \Phi(z^m)\right] dV(z).$$

We now examine an important formal property of momentum-energy complexes.

**Theorem 2.12.** *Let* $\{W_k{}^m(L)\}$ *denote the momentum-energy complex based on* L. *We have*

$$W_k{}^m(\alpha L_1 + \beta L_2) = \alpha W_k{}^m(L_1) + \beta W_k{}^m(L_2) \tag{2.138}$$

*for all constants* $\alpha, \beta$ *and for all lagrangian functions* $L_1, L_2$; $\{W_k{}^m(L)\}$ *acts linearly on the collection of all lagrangian functions.*

**Proof.** Let $\underset{1}{k}_a, \underset{1}{g}_a$ define the functional arguments of $L_1$, and let $\underset{2}{k}_a, \underset{2}{g}_a$ define the functional arguments of $L_2$. Let $\underset{1}{w}_{ak}{}^m$ be formed from $\underset{1}{g}_a$ and let $\underset{2}{w}_{ak}{}^m$ be formed from $\underset{2}{g}_a$. Equation (2.134) then gives

$$\frac{1}{\alpha} W_k{}^m(\alpha L_1) = \frac{\partial L_1}{\partial(\partial_m \Phi)} \partial_k \Phi + \int_{D^*} \underset{1}{w}_{ak}{}^m \frac{\partial L_1}{\partial \underset{1}{k}_a}(z)\, dV(z) - \delta_k{}^m L_1$$

$$\frac{1}{\beta} W_k{}^m(\beta L_2) = \frac{\partial L_2}{\partial(\partial_m \Phi)} \partial_k \Phi + \int_{D^*} \underset{2}{w}_{ak}{}^m \frac{\partial L_2}{\partial \underset{2}{k}_a}(z)\, dV(z) - \delta_k{}^m L_2$$

while

$$W_k{}^m(\alpha L_1 + \beta L_2) = \frac{\partial(\alpha L_1 + \beta L_2)}{\partial(\partial_m \Phi)} \partial_k \Phi - \delta_k{}^m(\alpha L_1 + \beta L_2)$$

$$+ \int_{D^*} \left[ \underset{1}{w}_{ak}{}^m \frac{\partial(\partial L_1 + \beta L_2)}{\partial \underset{1}{k}_a}(z) + \underset{2}{w}_{ak}{}^m \frac{\partial(L_1 + \beta L_2)}{\partial \underset{2}{k}_a}(z) \right] dV(z) .$$

However, $\partial L_1/\partial \underset{2}{k}_a = \partial L_2/\partial \underset{1}{k}_a \equiv 0$, and the result is established.

We have seen that the lagrangian function which leads to a given system of Euler equations is not unique. If L is such a lagrangian, then $L + A$ will have the same Euler equations as $L$ for any $A \in \mathfrak{N}(N)$. Hence, if we embed a given system of equations in a variational statement, we cannot actually distinguish between a continuum of lagrangian functions since the lagrangian function which accomplishes the embedding is determined only to within an equivalence class. Hence, it is impossible to make an *a priori* claim that either $L$ or $L + A$, for any $A \in \mathfrak{N}(N)$, is the lagrangian we seek. Now, with the characterization of $\mathfrak{N}(N)$ given in Sec. 2.3, the following result is easily demonstrated.

**Lemma 2.3.** *Let* $A \in \mathfrak{N}(N)$, *then* $W_k{}^m(A) \not\equiv 0$.

Lemma 2.3 is central to a problem which seems to have been totally ignored in the applications of variational principles in modern physics. Since we cannot distinguish between $L$ and $L + A$ for $A \in \mathfrak{N}(N)$ on an a priori basis in any problem of embedding, we may validly claim that *the momentum-energy complex associated with a given system of Euler equations is not unique*: if L embeds the given system of equations as Euler equations then we have the momentum-energy complexes

$$W_k{}^m(L),\ W_k{}^m(L + A_1),\ W_k{}^m(L + A_2),\ \ldots$$

for $A_1, A_2, \ldots \in \mathfrak{N}(N)$, and,

$$W_k{}^m(L + A) = W_k{}^m(L) + W_k{}^m(A) \not\equiv W_k{}^m(L)\,. \tag{2.139}$$

To add further complication, when (2.135) and $\{\mathcal{E}|L\}_{\Phi_\Lambda} = 0$ are used, we obtain

$$\frac{\partial W_k{}^m(L + A)}{\partial x^m} = \frac{\partial W_k{}^m(L)}{\partial x^m} + \frac{\partial W_k{}^m(A)}{\partial x^m} \tag{2.140}$$

on $S(L)$, and

$$\frac{\partial W_k{}^m(A)}{\partial x^m} = -\partial_k^* A - \int_{D^*}\left[\frac{\partial A}{\partial \hat{k}_a}(z^t)\,\partial_k^*\,\hat{g}_a^* - \frac{\partial A}{\partial \hat{k}_a}\frac{\partial \hat{g}_a}{\partial x^k}\right]dV(z) \tag{2.141}$$

where $\hat{k}_a$ is the functional argument in $A$ formed from $g_a$. We therefore have the following result.

**Theorem 2.13.** *If* $\{W_k{}^m(L)\}$ *is the mometum-energy complex formed on a lagrangian function* L *that embeds a given system* S *of equations as Euler equations, then* $\{W_k{}^m(L + A)\}$ *is also a momentum-energy complex for the system* S *for any* $A \in \mathfrak{N}(N)$. *Further, if* $\partial W_k{}^m(L)/\partial x^m \neq 0$, *then an* $A \in \mathfrak{N}(N)$ *can be found such that* $\partial W_k{}^m(L + A)/\partial x^m \neq 0$.

Basically, the converse problem is of more interest. Can we find an element $A$ of $\mathfrak{N}(N)$ so as to annihilate one of the $n$ functions $\partial W_k{}^m/\partial x^m$; i.e., can we find an $A \in \mathfrak{N}(N)$ such that

$$\frac{\partial W_k{}^m(L)}{\partial x^m} \neq 0 \quad , \quad \text{for } k \neq 1 \quad \text{and}$$

$$\frac{\partial W_k{}^m(L + A)}{\partial x^m} = 0 \quad , \quad \text{for } k = 1 \; ?$$

This remains an open and an important question. However a more demanding question is the following: can we find an $A \in \mathfrak{N}(N)$ such that if $\partial W_k{}^m(L)/\partial x^m \neq 0$, then $\partial W_k{}^m(L + A)/\partial x^m = 0$ for all $k$? The first question can probably be answered in the affirmative. The second question is not so obvious, although it would appear that the answer should be negative in general.

As an example, consider the equations

$$\partial_k \Phi = f_k \, .$$

These equations are imbedded by

$$L_1 = \Psi^k(\partial_k \Phi - f_k)$$

and by

$$2L_2 = \Psi^k(\partial_k \Phi - 2f_k) - \Phi \partial_k \Psi^k \, ,$$

and

$$W_k{}^m(L_1) = \Psi^m \partial_k \Phi - \delta_k{}^m(\Psi^t \partial_t \Phi - \Psi^t f_t) \, ,$$

$$2W_k{}^m(L_2) = \Psi^m \partial_k \Phi - \Phi \partial_k \Psi^m - \delta_k{}^m[\Psi^t(\partial_t \Phi - 2f_t) - \Phi \partial_t \Psi^t] \, .$$

However, for this case,

$$\frac{\partial W_k{}^m(L_1)}{\partial x^m} = \frac{\partial W_k{}^m(L_2)}{\partial x^m} = -\Psi^m \partial_k f_m \, .$$

since $(L_1 - L_2) \in \mathfrak{N}(2)$ is independent of $(x^m)$. Now,

$$A = \partial_t[h(x^m)\Psi^t(x^m)] = \Psi^t \partial_t h + h\partial_t \Psi^t \in \mathfrak{N}(2)$$

and

$$W_k{}^m(A) = h\partial_k \Psi^m - \delta_k{}^m(\Psi^t \partial_t h + h \partial_t \Psi^t) \, ,$$

$$\frac{\partial W_k{}^m(A)}{\partial x^m} = \Psi^t \partial_k \partial_t h + (\partial_t \Psi^t) \partial_k h \, ,$$

so that

$$W_k{}^m(L_1 + A) = \Psi^m \partial_k \Phi + h\partial_k \Psi^m - \delta_k{}^m(\Psi^t \partial_t \Phi + \Psi^t \partial_t h + h\partial_t \Psi^t - \Psi^t f_t),$$

$$\frac{\partial W_k{}^m(L_1 + A)}{\partial x^m} = -\Psi^m \partial_k f_m + \Psi^m \partial_k \partial_m h + (\partial_t \Psi^t)\partial_k h.$$

However, if $\left\{\mathcal{E} \mid L_1\right\}_{\Phi_1} \equiv \left\{\mathcal{E} \mid L_1 + A\right\}_{\Phi_1} = -\partial_k \Psi^k = 0$, then

$$\frac{\partial W_k{}^m(L_1 + A)}{\partial x^m} = \Psi^m(\partial_k \partial_m h - \partial_k f_m).$$

Hence, if $h(x^m)$ is any function such that $\partial_m h = f_m$, then

$$\frac{\partial W_k{}^m(L_1 + A)}{\partial x^m} = 0. \tag{2.142}$$

If the original system $\partial_m \Phi = f_m$ has a solution, then $\partial_k f_m - \partial_m f_k = 0$; and hence we can always find a function $h(x^m)$ such that $\partial_m h = f_m$. In this example, we achieve the desired result of finding an element of $\mathfrak{N}(2)$ such that (2.142) holds.

## 2.6. INVARIANCE OF THE EULER-EQUATIONS

We consider a general transformation $T$ on $\mathfrak{D}_1(D^*, N)$ to $\mathfrak{D}_1(D^*, N)$, defined by

$$T:_\Lambda\{\Phi\ (x^m)\} = \{f_\Lambda[\Psi_\Sigma(x^m), x^m]\}, \tag{2.143}$$

in which the functions $f_\Lambda$ are of class $C^2$ in their $N + n$ arguments and such that

$$\det\left(\frac{\partial f_\Lambda}{\partial \Psi_\Sigma}\right) \neq 0 \qquad \forall (x^m) \in D^*. \tag{2.144}$$

We then have

$$\partial_m \Phi_\Lambda = \frac{\partial f_\Lambda}{\partial \Psi_\Sigma} \partial_m \Psi_\Sigma + \partial^*_m f_\Lambda = \frac{\partial f_\Lambda}{\partial x^m}, \tag{2.145}$$

and hence

$$\frac{\partial(\partial_m \Phi_\Lambda)}{\partial(\partial_k \Psi_\Sigma)} = \delta_m{}^k \frac{\partial f_\Lambda}{\partial \Psi_\Sigma}, \qquad \frac{\partial(\partial f_\Lambda/\partial x^m)}{\partial \Psi_\Gamma} = \frac{\partial}{\partial x^m}\left(\frac{\partial f_\Lambda}{\partial \Psi_\Gamma}\right). \tag{2.146}$$

With the functions $g_a$ defined in Sec. 2.2, we define functions $\hat{g}_a$ by the relations

$$\hat{g}_a[x^m, z^m, \Psi_\Sigma(z^m), \partial_k \Psi_\Sigma(z^m)] \tag{2.147}$$
$$= g_a\left\{x^m, z^m, f_\Lambda[\Psi_\Sigma(z), z],\ \frac{\partial f_\Lambda}{\partial \Psi_\Sigma}(z)\,\partial_m \Psi_\Sigma(z) + \partial^*_m f_\Lambda(z)\right\}.$$

The $\hat{g}_a$ then define functionals $\tilde{k}_a(x, \Psi_\Sigma)$ by

$$\tilde{k}_a = \int_{D^*} \hat{g}_a dV(z)\ . \tag{2.148}$$

Similarly, define $\tilde{L}$ by the relations

$$\tilde{L}(x^m, \Psi_\Sigma, \partial_m \Psi_\Sigma, \tilde{k}_a) = L\left[x^m, f_\Lambda(\Psi_\Sigma, x),\ \frac{\partial f_\Lambda}{\partial \Psi_\Sigma}\partial_m \Psi_\Sigma + \partial^*_m f_\Lambda,\ k_a\right] \tag{2.149}$$

We now have, by (2.145) and (2.146),

$$\begin{aligned}\frac{\partial \tilde{L}}{\partial \Psi_\Gamma} &= \frac{\partial L}{\partial \Phi_\Lambda}\frac{\partial \Phi_\Lambda}{\partial \Psi_\Gamma} + \frac{\partial L}{\partial(\partial_m \Phi_\Lambda)}\frac{\partial(\partial_m \Phi_\Lambda)}{\partial \Psi_\Gamma}\\ &= \frac{\partial L}{\partial \Phi_\Lambda}\frac{\partial \Phi_\Lambda}{\partial \Psi_\Gamma} + \frac{\partial L}{\partial(\partial_m \Phi_\Lambda)}\frac{\partial(\partial f_\Lambda/\partial x^m)}{\partial \Psi_\Gamma}\end{aligned} \tag{2.150}$$

$$\frac{\partial \tilde{L}}{\partial(\partial_k \Psi_\Gamma)} = \frac{\partial L}{\partial \Phi_\Lambda}\frac{\partial \Phi_\Lambda}{\partial(\partial_k \Psi_\Gamma)} + \frac{\partial L}{\partial(\partial_m \Phi_\Lambda)}\frac{\partial(\partial_m \Phi_\Lambda)}{\partial(\partial_k \Psi_\Gamma)} = \frac{\partial L}{\partial(\partial_k \Phi_\Lambda)}\frac{\partial f_\Lambda}{\partial \Psi_\Gamma}\ . \tag{2.151}$$

Accordingly, we have

$$\{e|\tilde{L}\}_{\Psi_\Gamma}(x^m) = \frac{\partial L}{\partial \Phi_\Lambda}\frac{\partial f_\Lambda}{\partial \Psi_\Gamma} + \frac{\partial L}{\partial(\partial_m \Phi_\Lambda)}\frac{\partial(\partial f_\Lambda/\partial x^m)}{\partial \Psi_\Gamma} - \frac{\partial}{\partial x^k}\left(\frac{\partial L}{\partial(\partial_k \Phi_\Lambda)}\frac{\partial f_\Lambda}{\partial \Psi_\Gamma}\right),$$

and hence, when the second of (2.146) is used, we obtain

$$\{e|\tilde{L}\}_{\Psi_\Gamma}(x^m) = \{e|L\}_{\Phi_\Lambda}(x^m)\frac{\partial f_\Lambda}{\partial \Psi_\Gamma}\ . \tag{2.152}$$

If we start with the functionals $\tilde{k}_a$, then, in exactly the same fashion, we obtain

$$\left\{e \mid \tilde{g}_a^*\right\}_{\Psi_\Gamma}(x^m) = \left\{e \mid g_a^*\right\}_{\Phi_\Lambda}(x^m) \frac{\partial f_\Lambda[\Psi_\Sigma(x^m), x^m]}{\partial \Psi_\Gamma}, \tag{2.153}$$

while (2.149) gives

$$\frac{\partial \tilde{L}}{\partial \tilde{k}_a} = \frac{\partial L}{\partial k_a}. \tag{2.154}$$

We thus have the following result.

**Theorem 2.14.** *For any mapping*

$$T: \{\Phi_\Lambda(x^m)\} = \{f_\Lambda[\Psi_\Sigma(x^m), x^m\} \tag{2.155}$$

*of* $\mathfrak{D}_1(D^*, N)$ *to* $\mathfrak{D}_1(D^*, N)$ *such that the functions* $f_\Lambda$ *are class* $C^2$ *and* $\det(\partial f_\Lambda / \partial \Psi_\Sigma) \neq 0 \quad \forall x \in D^*$, *we have*

$$\left\{\mathcal{E} \mid \tilde{L}\right\}_{\Psi_\Gamma}(x^m) = \left\{\mathcal{E} \mid L\right\}_{\Phi_\Lambda}(x^m) \frac{\partial f_\Lambda}{\partial \Psi_\Gamma} \tag{2.156}$$

*for all* $\Phi_\Lambda \in \mathfrak{D}_1(D^*, N)$, *where* $\tilde{L}$ *is the function obtained by expressing* $\Phi$ *and its derivatives in terms of* $\Psi$ *and its derivatives.*

**Theorem 2.15.** *Let* $T$ *be a map of* $\mathfrak{D}_1(D^*, N)$ *to* $\mathfrak{D}_1(D^*, N)$ *which satisfies the conditions of Theorem 2.14 and let* $S(L)$ *denote the stationarization set of*

$$J[\{\Psi_\Gamma\}] = \int_{D^*} \tilde{L}\, dV(x),$$

*then* $T$ *maps* $S(L)$ *onto* $S(\tilde{L})$ *and* $T^{-1}$ *maps* $S(\tilde{L})$ *onto* $S(L)$

$$TS(L) = S(\tilde{L}), \quad T^{-1}S(\tilde{L}) = S(L). \tag{2.157}$$

**Proof.** Since $\det(\partial f_\Lambda / \partial \Psi_\Lambda) \neq 0 \quad \forall x \in D^*$, (2.156) shows that $\left\{\mathcal{E} \mid \tilde{L}\right\}_{\Psi_\Lambda}(x^m) = 0$ implies $\left\{\mathcal{E} \mid L\right\}_{\Psi_\Lambda}(x^m) = 0$ and $\left\{\mathcal{E} \mid L\right\}_{\Phi_\Lambda}(x^m) = 0$ implies $\left\{\mathcal{E} \mid \tilde{L}\right\}_{\Psi_\Gamma}(x^m) = 0$.

The above theorems are of central importance in a number of problems since we are free to perform mappings of the designated form without changing the stationarization of $J$.

*Problems*

2.1 Extend the results obtained in Sec. 1.9 to the case of several functions. Be careful to make all definitions and to define the functions $L^m$ and $s_b{}^m$ properly. You should end with $N$ formulas identical to (1.118) with $y(x^m)$ replaced by $\Phi_\Lambda(x^k)$.

2.2 Define the quantities $K^k$ and $M^k(s)$ by the relations

$$K^k = \sum_{s=0}^{r-1} \frac{M^k(s)}{s}, \qquad M^k(s) = W^{k\Lambda_1 i_1 \ldots \Lambda_s i_s} \partial_{i_1}\Phi_{\Lambda_1} \cdots \partial_{i_s}\Phi_{\Lambda_s},$$

in which $W$ is a function of $x$ and $\Phi$ and completely skew symmetric in the Latin indices and the Greek indices. Use equations (2.18)-(2.23) to show that any element of the null set of $\{\mathcal{E} \mid L\}_{\Phi_\Lambda}(x^m)$ can be written as $\partial_k K^k$ when $r = \min(n, N)$. Therefore, answer the question as to what divergences are variationally deletable.

2.3 Show by detailed calculation that (2.55) is the adjoint of (2.54).

2.4 Show by detailed calculation that (2.65) holds for any linear, second-order integrodifferential operator $\mathcal{O}$ when $L$ is given by (2.65) and when (2.63) defines $\tilde{\mathcal{O}}$.

2.5 Show that the diffusion equation

$$\frac{\partial \Phi}{\partial t} = k\left(\frac{\partial^2 \Phi}{\partial x^2} + \frac{\partial^2 \Phi}{\partial x^2} + \frac{\partial^2 \Phi}{\partial z^2}\right)$$

arises from the lagrangian function

$$L = -k\left\{\frac{\partial\Phi}{\partial x}\frac{\partial\Psi}{\partial x} + \frac{\partial\Phi}{\partial y}\frac{\partial\Psi}{\partial y} + \frac{\partial\Phi}{\partial z}\frac{\partial\Psi}{\partial z}\right\} - \frac{1}{2}\left(\Psi\frac{\partial\Phi}{\partial t} - \Phi\frac{\partial\Psi}{\partial t}\right).$$

Obtain this lagrangian function by the prescription given in the text (i.e., equations (2.63) and (2.64), and obtain $\{\mathcal{E} \mid L\}_\Psi(x^m)$. Show that $\Psi(t) = \Phi(-t)$.

2.6 Let $\Phi(x^m)$ be complex:

$$\Phi(x^m) = \Phi_1(x^m) + i\Phi_2(x^m),$$

then, since $\Phi_1(x^m)$ and $\Phi_2(x^m)$ are independent real functions, we may take $\Phi(x^m)$ and $\overline{\Phi}(x^m)$ as two independent functions with $\overline{\Phi}(x^m)$ as the complex conjugate of $\Phi(x^m)$. We thus consider a problem with $N = 2$, $\{\Phi_\Lambda(x)\} = \{\Phi(x), \overline{\Phi}(x)\}$. Show that the Schrodinger equation

$$-\frac{h^2}{2m}\left(\frac{\partial^2 \Phi}{\partial x^2} + \frac{\partial^2 \Phi}{\partial y^2} + \frac{\partial^2 \Phi}{\partial z^2}\right) + V(x^m)\Phi = ih\frac{\partial \Phi}{\partial t}$$

can be obtained from the lagrangian function

$$L = -\frac{h^2}{2m}\left\{\frac{\partial \Phi}{\partial x}\frac{\partial \overline{\Phi}}{\partial x} + \frac{\partial \Phi}{\partial y}\frac{\partial \overline{\Phi}}{\partial y} + \frac{\partial \Phi}{\partial z}\frac{\partial \overline{\Phi}}{\partial z}\right\} - V(x^m)\Phi\overline{\Phi} + \frac{1}{2}ih\left(\overline{\Phi}\frac{\partial \Phi}{\partial t} - \frac{\partial \overline{\Phi}}{\partial t}\Phi\right)$$

and obtain $\{\mathcal{E}|L\}_\Phi$, $\{\mathcal{E}|L\}_{\overline{\Phi}}$.

2.7 With

$$L = -\frac{h^2}{2m_0}\left\{\frac{\partial \Phi}{\partial x}\frac{\partial \overline{\Phi}}{\partial x} + \frac{\partial \Phi}{\partial y}\frac{\partial \overline{\Phi}}{\partial y} + \frac{\partial \Phi}{\partial z}\frac{\partial \overline{\Phi}}{\partial z}\right\} + \frac{h^2}{2m_0 c^2}\frac{\partial \Phi}{\partial t}\frac{\partial \overline{\Phi}}{\partial t} - \frac{m_0 c^2}{2}\overline{\Phi}\Phi$$

show that we obtain the equations for the charged scalar meson field;

$$\frac{\partial^2 \Phi}{\partial x^2} + \frac{\partial^2 \Phi}{\partial y^2} + \frac{\partial^2 \Phi}{\partial z^2} - \frac{\partial^2 \Phi}{\partial t^2}\frac{1}{c^2} = \frac{m_0{}^2 C^2}{h^2}\Phi$$

$$\frac{\partial^2 \overline{\Phi}}{\partial x^2} + \frac{\partial^2 \overline{\Phi}}{\partial y^2} + \frac{\partial^2 \overline{\Phi}}{\partial z^2} - \frac{1}{c^2}\frac{\partial^2 \overline{\Phi}}{\partial t^2} = \frac{m_0{}^2 C^2}{h^2}\overline{\Phi}.$$

2.8 Set $x^1 = x$, $x^2 = y$, $x^3 = z$, $x^4 = ict$, $A_1 = A_x$, $A_2 = A_y$, $A_3 = A_z$, $A_4 = iV$, where $(A_x, A_y, A_z)$ are the components of the vector potential and V is the scalar potential for the electromagnetic field. Use equations (2.63) and (2.64) to embed the equations for the charged scalar meson field inter acting with the electromagnetic field:

$$\left[\sum_{m=1}^{4}\left(\frac{\partial}{\partial x^m} - \frac{ie}{hc}A_m\right)^2 - \frac{m_0{}^2 c^2}{h^2}\right]\Phi = 0$$

$$\left[\sum_{m=1}^{4}\left(\frac{\partial}{\partial x^m} + \frac{ie}{hc}A_m\right)^2 - \frac{m_0^2 c^2}{h^2}\right]\overline{\Phi} = 0$$

where $e$ is the meson charge.

*Answer*: $L = -\dfrac{h^2}{2m_0}\left[\sum_{m=1}^{4}\left(\dfrac{\partial\overline{\Phi}}{\partial x^m} + \dfrac{ie}{hc}A_m\overline{\Phi}\right)\left(\dfrac{\partial\Phi}{\partial x^m} - \dfrac{ie}{\partial x^m}A_m\Phi\right)\right] - \dfrac{m_0 c^2}{2}\Phi\overline{\Phi}.$

2. Extend the theory of Sec. 1.9 to lagrangian functions and surface functions of several dependent functions. Show that

$\delta J[\{\Phi_\Lambda\}](L, L^k) = 0$, if and only if

i. $\{\mathcal{E} \mid L\}_{\Phi_\Lambda}(x^m) = 0$ in $D$, and

ii.
$$N_k(x^m)\left[\frac{\partial L}{\partial(\partial_k\Phi_\Lambda)} + \frac{\partial L^k}{\partial\Phi_\Lambda} + \int_{D*}\frac{\partial L}{\partial k_a}(z)\,\frac{\partial g_a^*}{\partial(\partial_k\Phi_\Lambda)}\,dV(z)\right.$$
$$\left. + \int_{\partial D}\frac{\partial L^k}{\partial s_b}(z)\,\frac{\partial s_b^{*t}}{\partial\Phi_\Lambda}\,dS_t(z)\right]_{\partial D} = 0\,,$$

or $\delta\Phi\,(x^m)\Big|_{\partial D} = 0$

2.10 Show that the natural boundary conditions are given by (ii) with $L^k \equiv 0$.

2.11 List all possible lagrangians with $N = 2$ for which the Euler equations are linear, and thereby extend to the case $N > 1$ the material covered in Sec. 1.7.

2.12 What is the most general lagrangian function of $\{\Phi_\Lambda\}$, $N > 1$, for which the Euler equations are self-adjoint?
Set $\mathcal{O}_{\Gamma\Lambda}\Phi_\Lambda = \{\mathcal{E} \mid L(\Phi)\}_\Phi$, where $L(\Phi)$ is given, and set $\widetilde{L} = \Psi_\Gamma\mathcal{O}_{\Gamma\Lambda}\Phi_\Lambda = \Psi_\Gamma\{\mathcal{E} \mid L(\Phi)\}_{\Phi_\Gamma}$. Then

$$\{\mathcal{E} \mid \widetilde{L}\}_{\Phi_\Gamma} = \mathcal{O}^+_{\Gamma\Lambda}\Psi_\Lambda = \left\{\mathcal{E} \mid \Psi_\Sigma\{\mathcal{E} \mid L(\Phi)\}_{\Phi_\Sigma}\right\}_{\Phi_\Gamma}.$$

If $\mathcal{O}_{\Gamma\Lambda} = \mathcal{O}^+_{\Gamma\Lambda}$, then we must have $\mathcal{O}^+_{\Gamma\Lambda}\Phi_\Lambda = \left\{\mathcal{E} \mid L(\Psi)\right\}_{\Psi_\Gamma}$. Conclude, therefore, that the most general $L$ which leads to self-adjoint Euler equations is a solution of

$$\left\{\mathcal{E} \mid \Psi_\Sigma \left\{\mathcal{E} \mid L(\Phi)\right\}_{\Phi_\Sigma}\right\}_{\Phi_\Gamma} = \left\{\mathcal{E} \mid L(\Psi)\right\}_{\Phi_\Gamma}.$$

2.13 Set

$$L = \Phi_1{}^2 - \Phi_2{}^2 + 3\Phi_1\Phi_2 + \frac{1}{2}\sum_{m=1}^{2} [(\partial_m \Phi_1)^2 + (\partial_m \Phi_2)^2 - 3\partial_m \Phi_1 \partial_m \Phi_2]$$

with $n = 2$. Compute the Euler equations for $L$ and for the lagrangian obtained under the mapping

$$\Phi_1 = \Psi_1 \cosh\sqrt{(x^1)^2 + (x^2)^2} + \Psi_2 \sinh\sqrt{(x^1) + (x^2)^2}$$

$$\Phi_2 = \Psi_1 \sinh\sqrt{(x^1)^2 + (x^2)^2} + \Psi_2 \cosh\sqrt{(x^1)^2 + (x^2)^2}$$

If $D^*$ is the unit disk, compare the norm $||\{\Phi_\Lambda\}||$ and the norm $||\{\Psi_\Lambda\}||$ when $\{\Psi_\Lambda\}$ and $\{\Phi_\Lambda\}$ are related by the above transformation.

# 3 Geometric Objects and Lie Derivatives

## 3.1. THE ARITHMETIC $n$-DIMENSIONAL MANIFOLD $A_n$

An ordered set of $n$ real or complex values of $n$ variables $y_{(k)}$, $k = 1, 2, \ldots, n$ is called an *arithmetic point*. The totality of all arithmetic points is an *arithmetic manifold* $A_n$ of $n$-dimensions. The $y_{(k)}$ are called the *components* of the arithmetic point. The point $P: y$ means the point with components $y_{(k)}$.

The set of all arithmetic points satisfying the inequalities

$$|y_{(k)} - y_{(k)0}| < z_k\ , \quad k = 1, 2, \ldots, n\ , \tag{3.1}$$

where $P: y_0$ is a given point of $A_n$ and $z_k$ are given positive numbers, is called a *box* of $A_n$ about $P: y_0$. A set of points $S$ of $A_n$ is said to be a *region* if and only if (1) every point of $S$ belongs to at least one box consisting of points of $S$; (2) for any two points of $S$ there exists a finite chain of boxes, all consisting of points of $S$, such that the first point lies in the first box and the second point lies in the last box and such that consecutive boxes have at least one point in common. Every box is a region and the whole $A_n$ is a region. Of course, not every region is a box. Every region is called a *neighborhood* of each of its points. For a neighborhood of $P: y$ we write $N(P)$ or $N(y_{(k)})$, as the situation dictates.

## 3.2. THE GEOMETRIC $n$-DIMENSIONAL MANIFOLD $X_n$

Let the elements of set $M$ be in one-to-one correspondence with the points of a region $R_0$ of $A_n$. It should be noted that we make no preassumptions concerning the nature of the elements of $M$. We call the one-to-one correspondence, $\tau$, between elements of $M$ and the points of $R_0$ a *coordinate system over* M. If a point $P:y$ corresponds to a certain element of $M$ under $\tau$, we call the $y_{(k)}$ *coordinates of this element* of M with respect to the one-to-one correspondence $\tau$ under consideration. If $y_{(k)}$ is considered in this way, we write

$$x^k = y_{(k)}$$

as the *coordinates on* $M$, that is $\tau: R_0 \longleftrightarrow M \,|\, x^k = y_{(k)}$. Each element of $M$ is then uniquely labeled by an ordered collection of $n$ real or complex numbers $(x^1, x^2, \ldots, x^n)$, and $M$ is said to be referred to the $(x)$-coordinate system.

Let $T$ denote a point transformation in $A_n$: any point $P:y$ in $A_n$ is moved to a new point $'P:'y$ under the action of $T$ with

$$'y_{(k)} = f_k(y_{(m)}) \,. \tag{3.2}$$

We require $T$ to be such that $f_k$ are functions of class $C^u$ in $N(y_0)$ for some $P:y_0$ in $A_n$ and that the functional determinant

$$\Delta \overset{\text{def}}{=} \operatorname{Det}\left(\frac{\partial f_k}{\partial y_{(m)}}\right) \tag{3.3}$$

is nonzero at $P:y_0$. There is then an inverse point transformation $T^{-1}$

$$y_{(k)} = F_k('y_{(m)}) \,, \qquad (F \cdot f = \text{identity}) \tag{3.4}$$

with the functions $F_k$ of $C^u$ at $'y_{(k)0} = f_k(y_{(m)0})$. There thus exists a neighborhood $N('y_0)$ and a neighborhood $N(y_0)$ for which (3.2) and (3.4) establish a one-to-one correspondence. Suppose that $N(y_0) \cap N('y_0) \neq \emptyset$, then there is a subset $\hat{R}$ of $M$ whose points are in a one-to-one correspondence with points of $N(y_0)$ in $N(y_0) \cap N('y_0)$ and in a one-to-one correspondence with the points of $N('y_0)$ in $N(y_0) \cap N('y_0)$. The latter correspondence is another coordinate system over $\hat{R}$. If we write

$$x^{k'} = 'y_{(k)} \,,$$

the $x^{k'}$ are new coordinates over $\hat{R}$, and $\hat{R}$ is said to be referred to the $(x')$-coordinate system.

It is important to note what occurs. In $A_n$ there is only one type of variable, and transformations mean point transformations: $y_{(k)} \to {'y}_{(k)}$, $N(y_0) \to N('y_0)$. In $M$ we have two types of variables, the coordinates $x^k$ and the coordinates $x^{k'}$. The transformations

$$x^{k'} = f_k(x^m)\,, \qquad x^k = F_k(x^{m'})\,, \qquad A_k^{k'} = \frac{\partial x^{k'}}{\partial x^k}\,. \tag{3.5}$$

in $\hat{R}$ are such that $\operatorname{Det}\left(A_k^{k'}\right) \neq 0$, and they transform coordinates only—the elements of $\hat{R}$ are unchanged. *A coordinate system over a set of elements* $\hat{R}$ *is a one-to-one correspondence between elements of* $\hat{R}$ *and the points of a region of* $A_n$. *Transformations of coordinates in* $\hat{R}$ *mean passing to other one-to-one correspondences between elements of* $\hat{R}$ *and the points of other regions of* $A_n$.

So far, we have only the set $R$ with coordinate systems over it and the possibility of introducing other coordinate systems by means of point transformations in $A_n$. Everything now depends on the choice of the point transformations. If we allow all transformations of class $C^u$ which map $N(y_0)$ onto $N(y_0)$, then the transformations form a group. If we allow the set $G_n{}^u$ of all transformations of points of $A_n$ such that each is of class $C^u$ in some region of $A_n$, but without the condition that the regions coincide, then the set $G_n{}^u$ does not form a group. Such a set of transformations is referred to as a *pseudogroup* since the composition of any two can be effected only if the domain of the second has a nonempty intersection with the range of the first.

The set $M$, provided with the pseudogroup $G_n{}^u$ and with all allowable coordinate systems (all coordinate systems that can be derived from a given system of coordinate coverings of subsets of $M$ by the action of $G_n{}^u$) is called a *general* $n$*-dimensional geometric manifold* $X_n{}^u$ of class $C^u$. We cannot use the notion of a box in $X_n{}^u$ because, contrary to $A_n$, there is no preferred coordinate system in $X_n{}^u$. Instead we use cells, a *cell* being defined as a set of points given by the inequalities

$$|x^k| < 1 \tag{3.6}$$

with respect to some allowable coordinate system. A point set $\hat{R}$ in $X_n{}^u$ is called a *region* if and only if there exists an allowable coordinate system determining a one-to-one correspondence between the points of $\hat{R}$ and the arithmetic points of a region $R$ of $A_n$. Every cell is a region, but not every region need be a cell. Every region $\hat{R}$ of $X_n{}^u$ is called a neighborhood of every one of the points of $\hat{R}$. For the *neighborhood* of an element of $X_n{}^u$ with coordinates $x^k$ with respect to the $(x)$-coordinate system, we write $N(x)$.

A function $p$ defined on a region $\hat{R}$ of $X_n{}^u$ is said to be of *class* $C^r$ if, with respect to some allowable coordinate system $(x)$ on $\hat{R}$, we have $p = f(x^k)$ and if $f(y_{(k)})$ is of class $C^r$ in the region $R$ corresponding to $\hat{R}$ by means of $(x)$. Careful note should be made of this definition in the case of $r > u$ because the pseudogroup on $X_n{}^u$ contains transformations of class $C^u$ only. Thus, a function on $\hat{R}$ which is of class $C^r$ for $r > u$ need not be of class $r (r > u)$ with respect to all coordinate systems; indeed, there will always be at least one coordinate system over $\hat{R}$ for which $f(y_{(k)})$ will be not more than of class $C^u$ on $R$ in $A_n$. To avoid this situation (continuity depending on the particular coordinate system under consideration) it is usual to consider functions of class $C^p$ for $p < u$ only.

## 3.3. GEOMETRIC OBJECTS ON $X_n{}^u$

An ordered collection of $N$ functions on $X_n{}^u$ which has the following properties is called a *geometric object of class* $p(\leq u)$:

1. In each coordinate system $(x)$, it has a well-determined set of $N$ components $\Phi_\Lambda(x)$, in which capital Greek indices run over the range $1, 2, \ldots, N$. (A more appropriate notation would be $\Phi_\Lambda(P:x)$ in which $P$ is a generic point of $X_n{}^u$, since $(x)$ in the notation $\Phi_\Lambda(x)$ denotes the *point* at which $\Phi_\Lambda$ is evaluated. The above notation is used for historical continuity.)
2. When a coordinate transformation $x^{k'} = f_k(x^m)$ is effected, the new components $\Phi_{\Lambda'}(x)$ of the object at the point $P: x(x')$ with respect to the new coordinate system $(x')$ can be represented as well determined functions of class $u - p$ of the old components $\Phi_\Lambda(x)$ at the point $P: x$, of the old coordinates $x^k$, of the functions $f_k$ (of the coordinates $x^{k'}$), and of the $s$-th partial derivatives of $f_k (1 \leq s \leq p \leq u)$; that is, the new components $\Phi_{\Lambda'}(x)$ of the object can be represented by equations of the form

$$\Phi_{\Lambda'}(x) = F_\Lambda(\Phi_\Sigma, x^k, f_k, \partial_{l_1} f_k, \ldots, \partial_{l_p \ldots l_1} f_k) \tag{3.7}$$

where $\partial_{l_p \ldots l_1} f_k = \partial^p f_k / \partial x^{l_p} \partial x^{l_{p-1}} \ldots \partial x^{l_1}$. (For the sake of simplicity we denote the right-hand side of (3.7) by

$$F_\Lambda(\Phi_\Sigma, x^k, x^{k'}) \text{ .)}$$

3. The functions $F_\Lambda(\Phi_\Sigma, x^k, x^{k'})$ have the group property:

$$F_\Lambda[F_\Sigma(\Omega, x, x'), x', x''] = F_\Lambda(\Omega, x, x'') , \tag{3.8}$$

$$F_\Lambda[F_\Sigma(\Omega, x, x'), x', x] = \Omega_\Lambda , \tag{3.9a}$$

and these imply

$$F_\Lambda(\Omega, x, x) = \Omega_\Lambda . \tag{3.9b}$$

When the functions $F_\Lambda(\Phi, x^k, x^{k'})$ contain only $\Phi_\Lambda(x^k)$ and the partial derivatives of the functions $f_k$ with respect to $x^m$, but not $x^k$ or $f_k$, the geometric object is said to be *differential*. When the functions $F_\Lambda(\Phi, x^k, x^{k'})$ are of the form

$$F_\Lambda(\Phi, x^k, x^{k'}) = F_\Lambda^{\ \Sigma}(x^k, x^{k'})\,\Phi_\Sigma(x^k) \tag{3.10}$$

the geometric object is said to be *linear homogeneous* and when the functions $F_\Lambda(\Phi, x^k, x^{k'})$ are of the form

$$F_\Lambda(\Phi, x^k, x^{k'}) = F_\Lambda^{\ \Sigma}(x^k, x^{k'})\,\Phi_\Sigma(x^k) + G_\Lambda(x^k, x^{k'}) \tag{3.11}$$

the geometric object is said to be *linear*.

Tensors are linear homogeneous differential geometric objects. A tensor of class $T(p, q)$ (contravariant order $p$ and convariant order $q$) has the transformation law

$$T^{i'_1 \ldots i'_p}_{j'_1 \ldots j'_q}(x^{k'}) = A_{i_1}^{\ i'_1} \ldots A_{i_p}^{\ i'_p} A_{j'_1}^{\ j_1} \ldots A_{j'_q}^{\ j_q}\, T^{i_1 \ldots i_p}_{j_1 \ldots j_q}(f_k^{-1}) . \tag{3.12}$$

An affine connection $L_{jk}^{\ \ i}$ with the transformation law

$$L_{j'k'}^{\ \ i'}(x^{k'}) = A_{j'}^{\ j} A_{k'}^{\ k}\left(A_i^{\ i'} L_{jk}^{\ \ i} - \partial_j A_k^{\ i'}\right) \tag{3.13}$$

is a linear differential geometric object.

When the components $\Psi_\Gamma(\Gamma = 1, \ldots, m)$ of a geometric object are functions of another geometric object $\Phi_\Omega$,

$$\Psi_\Gamma = \Psi_\Gamma(\Phi_\Omega) , \tag{3.14}$$

and the functional form of $\Psi_\Gamma(\Phi_\Omega)$ does not depend on the choice of the coordinate system, then $\Psi_\Gamma$ is said to be a *function of the geometric object* $\Phi_\Omega$.

We take an arbitrary field $\Phi_\Lambda(x^k)$ of a geometric object whose transformation law under a coordinate transformation $x \to (x')$ is

$$\Phi_\Lambda = F_\Lambda(\Phi_{\Sigma'}, x^{k'}, x^k) . \tag{3.15}$$

The law of transformation of $\partial_k \Phi_\Lambda$ is then given by

$$\partial_k \Phi_\Lambda = F_{k\Lambda}(\Phi_{\Lambda'}, \partial_{k'} \Phi_{\Lambda'}, x^{k'}, x^k) , \tag{3.16}$$

where the functions $F_{k\Lambda}$ are obtained from the $F_\Lambda$ by partial differentiation with respect to $x^k$. The quantities $\partial_k \Phi_\Lambda$ are not components of a geometric object, since the $F_{k\Lambda}$ depend on both $\Phi_\Lambda$ and $\partial_k \Phi_\Lambda$. However, $(\Phi_\Lambda, \partial_k \Phi_\Lambda)$ are components of a geometric object whose transformation law is given by

$$\begin{aligned} \Phi_\Lambda &= F_\Lambda(\Phi_{\Sigma'}, x^{k'}, x^k) , \\ \partial_k \Phi_\Lambda &= F_{k\Lambda}(\Phi_{\Lambda'}, \partial_{k'} \Phi_{\Lambda'}, x^{k'}, x^k) , \end{aligned} \tag{3.17}$$

or

$$\begin{aligned} \Phi_{\Lambda'} &= F_\Lambda(\Phi_\Sigma, x^k, x^{k'}) \\ \partial_{k'} \Phi_{\Lambda'} &= F_{k\Lambda}(\Phi_\Lambda, \partial_k \Phi_\Lambda, x^k, x^{k'}) . \end{aligned} \tag{3.18}$$

It is therefore convenient to consider the composite geometric object field ${}_\mu\Phi_\Lambda$, $(\mu = 0, 1, \ldots, n)$

$${}_0\Phi_\Lambda = \Phi_\Lambda , \qquad {}_k\Phi_\Lambda = \partial_k \Phi_\Lambda . \tag{3.19}$$

## 3.4. POINT TRANSFORMATIONS AND LIE DERIVATIVES IN $X_n^u$

Let $\bar{R}$ be a region of $X_n$ that is covered by the $(x)$-coordinate system, and let $T$ denote an analytic point transformation which moves the points of a subregion $\hat{R}_1$ of $\bar{R}$ to a subregion $\hat{R}_2$ of $\bar{R}$:

$$T : \ {}'x^k = f^k(x^m) , \qquad \mathrm{Det}(\partial_m f^k) \neq 0 . \tag{3.20}$$

Since $T$ is analytic, $T$ establishes a one-to-one correspondence between points in $\hat{R}_1$ and points in $\hat{R}_2$. During the point transformation, a point with coordinates $x^k$ in $\hat{R}_1$ is carried into a point with coordinates $'x^k$ in $\hat{R}_2$, where both $\hat{R}_1$ and $\hat{R}_2$ are referred to the same coordinate system $(x)$ of the enveloping region $\bar{R}$. Similarly, a point with coordinates $x^k + dx^k$ in $\hat{R}_1$ is carried into a point $'x^k + 'dx^k$ in $\hat{R}_2$.

Since $T$ carried a point $x^k$ in $\hat{R}_1$ into a point $'x^k$ in $\hat{R}_2$, the point transformation $T^{-1}$, inverse to $T$ ($T^{-1} \cdot T = T \cdot T^{-1}$ = identity transformation), carries the point $'x^k$ in $\hat{R}_2$ into the point $x^k$ in $\hat{R}_1$. With this inverse point transformation $T^{-1}: 'x \to x$, we can associate a coordinate transformation $(x) \to (x')$ by the requirement that

$$x^{k'} = 'x^k \tag{3.21}$$

(the new coordinates of the old point are equal to the old coordinates of the new point with the new point meaning the image of the old point under the point transformation and the new coordinates meaning the new coordinates under the coordinate transformation). A straightforward combination of (3.20) and (3.21) gives

$$x^{k'} = f_k(x^m)\ ; \quad \text{i.e.,} \quad f_k(x^m) \equiv f^k(x^m)\ . \tag{3.22}$$

This process, $(x) \to (x')$, is called the *dragging along of the coordinate system* $(x)$ *by the point transformation* $T^{-1}: 'x \to x$, and the coordinate system $(x')$ is called the *coordinate system dragged along by* $T^{-1}$.

Let $\Phi_\Lambda(x)$ be a geometric object field over $\bar{R} \subset X_n{}^u$, then to each point in $\bar{R}$ referred to the $(x)$-coordinate system, we have $N$ values associated by the evaluation of $\Phi_\Lambda(x)$ at the point under study. Since both $\hat{R}_1$ and $\hat{R}_2$ are contained in $\bar{R}$, the values of this geometric object field at $'x \in \hat{R}_2$, in which $T: x^k \to 'x^k$, are known and called $\Phi_\Lambda('x)$. We now define a new field $'\Phi_\Lambda(x)$ of geometric objects over $\hat{R}_1$ by the following conditions:

1. The functions $F_\Lambda('\Phi, x^k, x^{k'})$ defining the transformation law of $'\Phi_\Lambda$ are the same functions as those defining the transformation law for $\Phi_\Lambda$.
2. The values of the new field at the old point in $\hat{R}_1$ with respect to the coordinate system $(x')$ dragged along by $T^{-1}$ are numerically equal to the values of the field at the image of $x^k$ in $\hat{R}_2$ under $T$; that is,

$$'\Phi_{\Lambda'}(x) = \Phi_\Lambda('x)\ . \tag{3.23}$$

This process $\Phi_\Lambda \to '\Phi_\Lambda$ is called the *dragging along of the field* $\Phi_\Lambda$ *by the point transformation* $T^{-1}$, and the field $'\Phi_\Lambda$ is called the *dragged along field.* Occasionally, it is said that $T^{-1}$ *deforms* the field $\Phi_\Lambda$ into the field $'\Phi_\Lambda$ and $'\Phi_\Lambda$ is referred to as the *deformed field.*

Consider now an infinitesimal point transformation in $X_n$ given by

$$T: \ 'x^k = x^k + v^k(x^m)\,t + o(t) \ . \tag{3.24}$$

The coordinate system dragged along by $T^{-1}$ is then given by

$$x^{k'} = 'x^k = x^k + v^k(x^m)\,t + o(t) \ . \tag{3.25}$$

Suppose that we have a geometric object field $\Phi_\Lambda(x)$ with the transformation law $\Phi_{\Lambda'}(x) = F_\Lambda(\Phi_\Omega, x^k, x^{k'})$ in $X_n$. From the definition of the field dragged along by the point transformation $T^{-1}$, we have

$$'\Phi_{\Lambda'}(x) \equiv \Phi_\Lambda('x) \ . \tag{3.26}$$

Now, $'\Phi_\Lambda(x)$ is a geometric object field with the same transformation law as $\Phi_\Lambda(x)$, and hence the relation between $'\Phi_{\Lambda'}(x)$ and $'\Phi_\Lambda(x)$ is given by

$$'\Phi_{\Lambda'}(x) = F_\Lambda('\Phi_\Omega, x^k, x^{k'}) \ . \tag{3.27}$$

The *Lie differential* $\underset{v}{\pounds}\Phi_\Lambda dt$ and the *Lie derivative* $\underset{v}{\pounds}\Phi_\Lambda$ of the geometric object $\Phi_\Lambda$ with respect to the point transformation (3.24) are defined by

$$\underset{v}{\pounds}\Phi_\Lambda dt = '\Phi_\Lambda(x) - \Phi_\Lambda(x) + o(dt) \tag{3.28}$$

and

$$\underset{v}{\pounds}\Phi_\Lambda = \lim_{t\to 0} \frac{'\Phi_\Lambda(x) - \Phi_\Lambda(x)}{t} \ . \tag{3.29}$$

Let $\Phi_\Lambda(x)$ be a geometric object field of class $p \leq u$ on $X_n{}^u$. From (3.26) and (3.27), we have

$$\Phi_\Lambda('x) = F_\Lambda('\Phi, x^k, x^{k'}) \ . \tag{3.30}$$

However, from

$$'x^k = f^k(x^m) = x^k + v^k(x^m)\,t + o(t)$$

we easily obtain

$$\partial_m f^k = \delta_m{}^k + \partial_m v^k t + o(t)$$

$$\partial_{m_2} \partial_{m_1} f^k = \partial_{m_2} \partial_{m_1} v^k t + o(t) \overset{\text{def}}{=} \partial_{m_2 m_1} v^k t + o(t)$$

$$\partial_{m_p \dots m_1} f^k = \partial_{m_p \dots m_1} v^k t + o(t) \,.$$

Substituting into (3.30) (into (3.28)), we then have

$$\Phi_\Lambda(x) + v^k \partial_k \Phi_\Lambda(x) t + o(t) = \Phi_\Lambda('x)$$

$$= {}'\Phi_\Lambda(x) + \sum_{s=0}^{p} F_k^{m(s)}(\Phi, x) \partial_{m(s)} v^k t + o(t) \,, \tag{3.31}$$

where

$$\partial_{m(0)} v^k = v^k \,, \qquad \partial_{m(s)} v^k = \partial_{m_s \dots m_1} v^k$$

$$F_{k\Lambda}^{m(0)} = \left[\frac{\partial F_\Lambda}{\partial f^k}\right] , \qquad F_{k\Lambda}^{m(s)} = \left[\frac{\partial F}{\partial\left(\partial_{m_s \dots m_1} f^k\right)}\right] \tag{3.32}$$

and the $[\dots]$ denotes evaluation of the enclosed expressions for

$$f^k = x^k \,, \quad \partial_m f^k = \delta_m{}^k \,, \quad \partial_{m_2 m_1} f^k = 0 \,, \dots , \quad \partial_{m_p \dots m_1} f^k = 0 \,. \tag{3.33}$$

When (3.32) is solved for $'\Phi_\Lambda(x)$ and the results are substituted into (3.29), we obtain the following explicit evaluation of $\underset{v}{£}\, \Phi_\Lambda$:

$$\underset{v}{£}\, \Phi_\Lambda(x) = v^k \partial_k \Phi_\Lambda - \sum_{s=0}^{p} F_k^{m(s)}(\Phi, x) \partial_{m(s)} v^k \,. \tag{3.34}$$

With this result, the statement (3.31) assumes the equivalent form

$$\Phi_\Lambda('x) - {}'\Phi_\Lambda(x) = \underset{v}{D}\, \Phi_\Lambda t + o(t) \tag{3.35}$$

where

$$\underset{v}{D}\, \Phi_\Lambda = v^k \partial_k \Phi_\Lambda - \underset{v}{£}\, \Phi_\Lambda \,. \tag{3.36}$$

The operator defined by (3.36) is of interest in its own right. If we have only coordinate transformations

$$x^{k'} = x^k + v^k(x^m)\,t + o(t)\,,$$

equations (3.8) and the above considerations give

$$\Phi_{\Lambda'}(x) - \Phi_\Lambda(x) = \underset{v}{D}\,\Phi_\Lambda(x)\,t + o(t)\,, \tag{3.37}$$

and hence, by (3.35) and (3.37)

$$\Phi_\Lambda('x) - \Phi_{\Lambda'}(x) = {'\Phi_\Lambda}(x) - \Phi_\Lambda(x) + o(t)\,. \tag{3.38}$$

Under a general coordinate transformation $(x) \to (x')$, ${'\Phi_\Lambda}(x)$ is transformed into ${'\Phi_{\Lambda'}} = F_\Lambda('\Phi, x^k, x^{k'})$ and $\Phi_\Lambda(x)$ is transformed into $\Phi_{\Lambda'} = F_\Lambda(\Phi, x^k, x^{k'})$ , as follows from the definition of a geometric object and condition 1 of the definition of ${'\Phi_\Lambda}(x)$. We thus obtain

$$\underset{v}{\pounds}\,\Phi_{\Lambda'} = \frac{\partial F_\Lambda}{\partial \Phi_\Sigma}\,\underset{v}{\pounds}\,\Phi_\Sigma\,. \tag{3.39}$$

This equation gives the transformation law for the Lie derivative of $\Phi_\Lambda$ during a coordinate transformation $(x) \to (x')$.

Since the partial derivatives $\partial F_\Lambda/\partial\Phi_\Sigma$ contain $\Phi_\Lambda$ in general, the Lie derivative of a general geometric object is not necessarily a geometric object. This follows from the fact that $\underset{v}{\pounds}\,\Phi_{\Lambda'}$ will depend on $\Phi_\Lambda(x)$ as well as $\underset{v}{\pounds}\,\Phi_\Lambda(x)$ if $\partial F_\Lambda/\partial\Phi_\Sigma$ contains $\Phi_\Lambda$. In general, we have that the composite object $(\Phi_\Lambda, \underset{v}{\pounds}\,\Phi_\Lambda)$ is always a geometric object, but $\underset{v}{\pounds}\,\Phi_\Lambda$ by itself is generally not a geometric object. It is evident that the Lie derivative of a geometric object is a geometric object if and only if the partial derivatives $\partial F_\Lambda/\partial\Phi_\Sigma$ do not contain $\Phi$ ; that is, if and only if $F_\Lambda(\Phi, x^k, x^{k'})$ has the form

$$F_\Lambda(\Phi, x^k, x^{k'}) = F_\Lambda{}^\Sigma(x^k, x^{k'})\,\Phi_\Sigma + G_\Lambda(x^k, x^{k'})\,. \tag{3.40}$$

From this we conclude that *a necessary and sufficient condition that the Lie derivative of a geometric object be a geometric object is that the original geometric object be linear in which case the Lie derivative is a linear homogeneous geometric object.* For a linear geometric object, we have

$$\underset{v}{\pounds}\,\Phi_\Lambda = v^k\partial_k\Phi_\Lambda - \sum_{s=0}^{p}\left[F^{m_{(s)}\Sigma}_{k_\Lambda}(x)\,\Phi_\Sigma + G^{m_{(s)}}_{k_\Lambda}(x)\right]\partial_{m_{(s)}}v^k\,. \tag{3.41}$$

As an immediate consequence of (3.41), we have

$$\underset{v}{£}(\Phi_\Lambda \pm \Psi_\Lambda) = \underset{v}{£}\Phi_\Lambda \pm \underset{v}{£}\Psi_\Lambda \tag{3.42}$$

*if* $\Phi_\Lambda$ *and* $\Psi_\Lambda$ *are linear homogeneous and have the same* $F^\Sigma_\Lambda(x^k, x^{k'})$. *If* $\Phi_\Lambda$ *and* $\Psi_\Lambda$ *are only linear and have the same* $F^\Sigma_\Lambda(x, x')$ *and* $G_\Lambda(x')$, *then*

$$\underset{v}{£}(\Phi_\Lambda - \Psi_\Lambda) = \underset{v}{£}\Phi_\Lambda - \underset{v}{£}\Psi_\Lambda \tag{3.43}$$

$$\underset{v}{£}(\Phi_\Lambda + \Psi_\Lambda) = \underset{v}{£}\Phi_\Lambda + \underset{v}{£}\Psi_\Lambda - G^{m(s)}_{k\Lambda}\,\partial_{m_{(s)}} v^k . \tag{3.44}$$

*If* $\Phi_\Lambda$ *and* $\Psi_\Lambda$ *are linear homogeneous and* $\Phi_\Lambda \times \Psi_\Lambda$ *is linear homogeneous, then*

$$\underset{v}{£}(\Phi_\Lambda \times \Psi_\Sigma) = (\underset{v}{£}\Phi_\Lambda) \times \Psi_\Sigma + \Phi_\Lambda \times \underset{v}{£}\Psi_\Sigma . \tag{3.45}$$

One additional result will be needed repeatedly. If $\Phi_\Lambda$ is a geometric object field, then

$$\begin{aligned} \Phi_\Lambda &= F_\Lambda[\Phi_{\Lambda'}, x', x] \\ \partial_m \Phi_\Lambda &= F_{m\Lambda}[\Phi_{\Sigma'}, \partial_{k'}\Phi_{\Lambda'}, x^{k'}, x^k] \end{aligned} \tag{3.46}$$

and consequently, by (3.26) and (3.27)

$$\left\{ \begin{aligned} {}'\Phi_\Lambda &= F_\Lambda[\Phi_\Sigma('x), x^{k'}, x^k] \\ \partial_m {}'\Phi_\Lambda &= F_{m\Lambda}[\Phi_\Sigma('x), \partial_k \Phi_\Sigma('x), x^{k'}, x^k] . \end{aligned} \right. \tag{3.47}$$

Therefore, by (3.29) and the fact that $(\Phi_\Lambda, \partial_m \Phi_\Lambda)$ is a geometric object, we have

$$\underset{v}{£}(\Phi_\Lambda, \partial_m \Phi_\Lambda) = (\underset{v}{£}\Phi_\Lambda, \partial_m \underset{v}{£}\Phi_\Lambda) . \tag{3.48}$$

On the other hand, (3.47) can be written as

$$\begin{aligned} {}'\Phi_\Lambda &= F_\Lambda[\Phi_\Sigma('x), x^{k'}, x^k] , \\ {}'(\partial_m \Phi_\Lambda) &= F_{m\Lambda}[\Phi_\Sigma('x), \partial_m \Phi_\Sigma('x), x^{k'}, x^k] , \end{aligned} \tag{3.49}$$

and consequently

$$\underset{v}{£}(\Phi_\Lambda, \partial_m \Phi_\Lambda) = (\underset{v}{£}\Phi_\Lambda, \underset{v}{£}\partial_m \Phi_\Lambda) . \tag{3.50}$$

Comparing (3.48) and (3.50), we have

$$\underset{v}{\pounds}\,\partial_m \Phi_\Lambda = \partial_m \underset{v}{\pounds}\,\Phi_\Lambda : \tag{3.51}$$

*Lie differentiation and partial differentiation are commutative operators.*

Of the several equivalent methods of defining the Lie derivatives, the one introduced above is the most useful from the standpoint of calculation. We shall now give another development which is intuitatively simple, but which is not the most convenient form for actual calculations.

We start as before with a point transformation $T$ that maps a subregion $\hat{R}_1 \subset \bar{R}$ onto a subregion $\hat{R}_2 \subset \bar{R}$:

$$T:\ {'x^k} = f^k(x^m), \quad \det(\partial_m f^k) \neq 0 . \tag{3.52}$$

The functions $f^k$ are assumed to be analytic in the region considered. In the previous consideration, we based our analysis on the inverse transformation $T^{-1}$. Here we base the considerations on $T$ itself. We introduce a new coordinate system by the requirement that "the new coordinates of the new point are numerically equal to the old coordinates of the old point." Hence, if

$$x^k = F^k('x^m) \tag{3.53}$$

is the inverse of (3.52) ($F \cdot f$ = identity), we must have

$${'x^{k'}} = x^k = F^k('x^m) . \tag{3.54}$$

Accordingly, the coordinate transformation from $(x)$ to $(x')$ is given by

$$x^{k'} = F^k(x^m), \quad x^k = f^k(x^{m'}) . \tag{3.55}$$

This process is called *the dragging along of the coordinate system* $(x)$ *by the point transformation* $T$ (as opposed to dragging along by $T^{-1}$).

The following example gives a simple interpretation of the coordinate system dragged along by $T$. Suppose that the region considered is filled with gelatine and that the coordinate net associated with the $(x)$-coordinate system is painted red (a coordinate net is the set of coordinate lines that define the locations of points with respect to the $(x)$-coordinate system). If every point of the gelatine is moved about in the region by the point transformation $T$, the red curves in their new positions are the coordinate net of the coordinate system $('x)$ that is dragged along by $T$. Thus, a dragged along coordinate system is seen to be what is commonly referred to as a system of convected coordinates if the point transformation $T$ is thought of as describing the flow of a fluid (i.e., lagrangian coordinates in the classical terminology).

Let $\Phi_\Lambda(x^m)$ be a geometric object field. We define a new field ${}^m\Phi_\Lambda(x^m)$ by the requirement that the "the new field at the new point referred to the new coordinates is numerically equal to the old field at the old point referred to the old coordinates." We accordingly have

$$ {}^m\Phi_{\Lambda'}('x^k) = \Phi_\Lambda(x^k) . \tag{3.56}$$

This process is called *the dragging along of a field by the point transformation* $T$ (as opposed to the dragging along by $T^{-1}$).

If the transformation $T$ is infinitesimal:

$$ 'x^k = x^k + v^k(x^m)\,dt + o(dt) , \tag{3.57}$$

then it is customary to refer to the process of dragging along by $T$ as *dragging along over* $v^k(x^m)\,dt$. The process of dragging along an infinitesimal contravariant vector field $U^k(x^m)\,ds$ over $v^k(x^m)\,dt$ is illustrated in Figure 1.

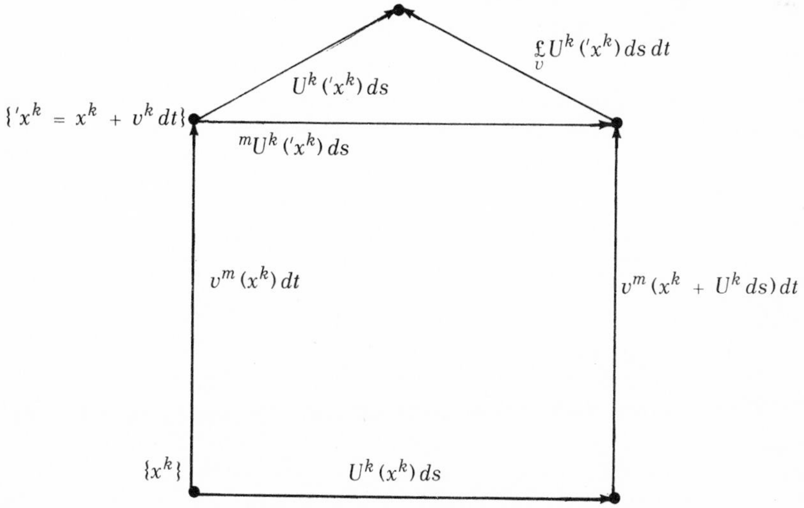


Fig. 1. The Lie Derivative.

The vector ${}^mU^k$ at ${}'x^k = x^k + v^k\,dt$ results from the displacement of the beginning and end points of the vector $U^k\,ds$ at $x^k$ by the field $v^k\,dt$. In general, ${}^mU^k(x^m + v^m\,dt)$ and $U^k(x^m + v^m\,dt)$ do not coincide, and thus a pentagon results. Even though the fifth side of the pentagon is the Lie derivative, it should not be construed that Lie derivative is a fiendish object.

Restricting our attention to the case of infinitesimal point transformations (3.57), we define the Lie derivative and the Lie differential by the relations

$$\underset{v}{\pounds}\,\Phi_\Lambda(x:) = \lim_{t\to 0} \frac{\Phi_\Lambda(x + vt) - {}^m\Phi_\Lambda(x + vt)}{t} \tag{3.58}$$

and

$$\Phi_\Lambda(x + v\,dt) - {}^m\Phi_\Lambda(x + v\,dt) = \underset{v}{\pounds}\,\Phi_\Lambda(x)\,dt + o(dt)\,. \tag{3.59}$$

Now,

$$\Phi_\Lambda(x + v\,dt) = \Phi_\Lambda(x) + v^k\,\partial_k\,\Phi_\Lambda(x)\,dt + o(dt) \tag{3.60}$$

and

$${}^m\Phi_\Lambda(x + v\,dt) = \Phi_\Lambda(x) + {}^m d\Phi_\Lambda(x) + o(dt) \tag{3.61}$$

where ${}^m d\Phi_\Lambda$ is the increment due to dragging $\Phi_\Lambda$ over $v\,dt$. On combining (3.59) - (3.61), we accordingly have

$$\underset{v}{\pounds}\,\Phi_\Lambda(x)\,dt = v^k\,\partial_k\,\Phi_\Lambda\,dt - {}^m d\Phi_\Lambda(x) + o(dt)\,, \tag{3.62}$$

and hence

$${}^m d\Phi_\Lambda(x) = (v^k\,\partial_k\,\Phi_\Lambda - \underset{v}{\pounds}\,\Phi_\Lambda)(x)\,dt\,. \tag{3.63}$$

A comparison of (3.63) and (3.36) then gives

$${}^m d\Phi_\Lambda(x) = \underset{v}{D}\,\Phi_\Lambda(x)\,dt\,, \tag{3.64}$$

since it is easily shown that the two definitions of the Lie derivative are equivalent (that is, $\Phi_\Lambda({}'x) = {}'\Phi_{\Lambda'}(x)$, ${}^m\Phi_{\Lambda'}({}'x) = \Phi_\Lambda(x)$, etc.). We thus have

$${}^m\Phi_\Lambda({}'x) = \Phi_\Lambda(x) + \underset{v}{D}\,\Phi_\Lambda(x)\,t + o(t)\,. \tag{3.65}$$

It thus follows that *the variation in* $\Phi_\Lambda(x)$, *that is induced by a point transformation* $T$, *is given by*

$$ {}^m\Phi_\Lambda('x) - \Phi_\Lambda(x) = \underset{v}{D}\Phi_\Lambda(x)\,dt\,. \tag{3.66}$$

This result will allow us to solve an otherwise difficult aspect of invariance theory in Chap. 4 since $-\underset{v}{D}\Phi_\Lambda(x)\,dt$ is now known to be the variation induced in $\Phi_\Lambda$ by the point transformation $T^{-1}$.

## 3.5. LINEAR GEOMETRIC OBJECTS

The most often encountered geometric objects are linear, and thus this class will be studied in greater detail. Let $\Phi_\Lambda(x)$ be a linear geometric object of class $p$, that is, a geometric object with the transformation law

$$\Phi_{\Lambda'} = F_\Lambda{}^\Sigma(x^k, x^{k'})\,\Phi_\Sigma + G_\Lambda(x^k, x^{k'})\,. \tag{3.67}$$

Since the transformation functions of a geometric object must satisfy the conditions given by (3.8) through (3.9), the functions $F_\Lambda{}^\Sigma(x^k, x^{k'})$ and $G_\Lambda(x^k, x^{k'})$ appearing in (3.67) must satisfy the relations

$$F_\Lambda{}^\Sigma(x^{k'}, x^{k''})\,F_\Sigma{}^\Gamma(x^k, x^{k'}) = F_\Lambda{}^\Gamma(x^k, x^{k''}) \tag{3.68}$$

$$F_\Lambda{}^\Sigma(x^{k'}, x^{k''})\,G_\Sigma(x^k, x^{k'}) + G_\Lambda(x^{k'}, x^{k''}) = G_\Lambda(x^k, x^{k''}) \tag{3.69}$$

and

$$F_\Lambda{}^\Sigma(x^k, x^k) = \delta_\Lambda{}^\Sigma, \quad G_\Lambda(x^k, x^k) = 0\,. \tag{3.70}$$

Consider two infinitesimal point transformations

$$'x^k = x^k + \underset{1}{v}{}^k(x)\,t + o(t)\,, \quad ''x^k = {}'x^k + \underset{2}{v}{}^k('x)\,u + o(u) \tag{3.71}$$

and the dragged along coordinate transformations

$$x^{k'} = x^k + \underset{1}{v}{}^k(x)\,t + o(t)\,, \quad x^{k''} = x^{k'} + \underset{2}{v}{}^k(x')\,u + o(u)\,. \tag{3.72}$$

When the first of (3.72) is substituted into the second, we then obtain the composite form

$$x^{k''} = x^k + \underset{1}{v}{}^k(x)\,t + \underset{2}{v}{}^k(x)\,u + \underset{1}{v}{}^m\partial_m\underset{2}{v}{}^k(x)\,ut + \cdots. \tag{3.73}$$

When (3.71) are used to evaluate the left-hand side of (3.68) to within second-order terms in $t$ and $u$ and (3.73) is used to evaluate the right-hand side of (3.68) to within second-order terms in $t$ and $u$, the identical satisfaction of (3.68) for all such infinitesimal transformations allows us to equate the coefficients of the powers of $t$ and $u$ in the resulting expressions. When (3.70) is used, a lengthy but straightforward calculation leads to the following result:

$$\{F, \underset{2}{v}\}_\Lambda{}^\Sigma \{F, \underset{1}{v}\}_\Sigma{}^\Gamma + \underset{1}{v}^m \partial_m \{F, \underset{2}{v}\}_\Lambda{}^\Gamma = \{F, \underset{1}{v}^m \partial_m \underset{2}{v}\}_\Lambda{}^\Gamma + \{F, \underset{2}{v}, \underset{1}{v}\}_\Lambda{}^\Gamma \tag{3.74}$$

where

$$\begin{aligned} \{F, v\}_\Lambda{}^\Sigma &\overset{\text{def}}{=} \sum_{s=0}^{p} F_{k\Lambda}{}^{m(s)\Sigma} \partial_{m(s)} v^k , \\ \{F, \underset{2}{v}, \underset{1}{v}\}_\Lambda{}^\Sigma &= \sum_{s,t=0}^{p} F_{kj\Lambda}^{m(s)m(t)\Sigma} (\partial_{m(s)} \underset{2}{v}^k)(\partial_{m(t)} \underset{1}{v}^j) . \end{aligned} \tag{3.75}$$

Similarly, the conditions (3.69) lead to the results

$$\{F, \underset{2}{v}\}_\Lambda{}^\Sigma \{G, \underset{1}{v}\}_\Sigma + \underset{1}{v}^m \partial_m \{G, \underset{2}{v}\}_\Lambda = \{G, \underset{1}{v}^m \partial_m \underset{2}{v}\}_\Lambda + \{G, \underset{2}{v}, \underset{1}{v}\}_\Lambda , \tag{3.76}$$

where

$$\begin{aligned} \{G, v\}_\Lambda &= \sum_{s=0}^{p} G_{k\Lambda}{}^{m(s)} \partial_{m_s} v^k \\ \{G, \underset{2}{v}, \underset{1}{v}\} &= \sum_{s,t=0}^{p} G_{kj\Lambda}^{m(s)m(t)} (\partial_{m(s)} \underset{2}{v}^k)(\partial_{m(t)} \underset{1}{v}^i) . \end{aligned} \tag{3.77}$$

In these formulas we have used

$$F_{kj\Lambda}^{m(s)m(t)\Sigma} = \left[ \frac{\partial^2 F_\Lambda{}^\Sigma}{\partial(\partial_{m(s)} f^k)\, \partial(\partial_{m(t)} f^j)} \right], \quad \text{etc.} \tag{3.78}$$

We now consider $r$ infinitesimal point transformations

$$'x^k = x^k + \underset{a}{v}^k (x^m)\, t + o(t) . \quad a = 1, \ldots, r . \tag{3.79}$$

For any two of the vectors $v(x^m)$, for example, $\underset{b}{v}(x^m)$ and $\underset{c}{v}(x^m)$, formulas (3.74) and (3.75) give

$$\{F, \underset{c}{v}\}_\Lambda{}^\Sigma \{F, \underset{b}{v}\}_\Sigma{}^\Gamma - \{F, \underset{b}{v}\}_\Lambda{}^\Sigma \{F, \underset{c}{v}\}_\Sigma{}^\Gamma - \underset{c}{v}{}^m \partial_m \{F, \underset{b}{v}\}_\Lambda{}^\Gamma + \underset{b}{v}{}^m \partial_m \{F, \underset{c}{v}\}_\Lambda{}^\Gamma$$
$$= -\{F, \underset{c}{v}{}^m \partial_m \underset{b}{v} - \underset{b}{v}{}^m \partial_m \underset{c}{v}\}_\Lambda{}^\Gamma, \tag{3.80}$$

$$\{F, \underset{c}{v}\}_\Lambda{}^\Sigma \{G, \underset{b}{v}\}_\Sigma - \{F, \underset{b}{v}\}_\Lambda{}^\Sigma \{G, \underset{c}{v}\}_\Sigma - \underset{c}{v}{}^m \partial_m \{G, \underset{b}{v}\}_\Lambda + \underset{b}{v}{}^m \partial_m \{G, \underset{c}{v}\}$$
$$= -\{G, \underset{c}{v}{}^m \partial_m \underset{b}{v} - \underset{b}{v}{}^m \partial_m \underset{c}{v}\}_\Lambda. \tag{3.81}$$

The Lie derivatives of the linear geometric object $\Phi_\Lambda$ with respect to $\underset{b}{v}(x^m)$ are now given by

$$\underset{b}{\pounds}\Phi_\Lambda = \underset{b}{v}{}^m \partial_m \Phi_\Lambda - (\{F, \underset{b}{v}\}_\Lambda{}^\Sigma \Phi_\Sigma + \{G, \underset{b}{v}\}_\Lambda), \tag{3.82}$$

where $\underset{b}{\pounds}\Phi_\Lambda$ represents the Lie derivative with respect to $\underset{b}{v}(x^m)$. Since $\Phi_\Lambda$ is linear, $\underset{b}{\pounds}\Phi_\Lambda$ is a linear homogeneous geometric object with the transformation law

$$\underset{b}{\pounds}\Phi_{\Lambda'} = F_\Lambda{}^\Sigma(x^k, x^{k'})\underset{b}{\pounds}\Phi_\Sigma. \tag{3.83}$$

We thus obtain

$$\underset{c}{\pounds}\underset{b}{\pounds}\Phi_\Lambda = \underset{c}{v}{}^k \Big[\partial_k \underset{b}{v}{}^m \partial_m \Phi_\Lambda + \underset{b}{v}{}^m \partial_k \partial_m \Phi_\Lambda - (\partial_k \{F, \underset{b}{v}\}_\Lambda{}^\Sigma)\Phi_\Sigma - \{F, \underset{b}{v}\}_\Lambda{}^\Sigma \partial_k \Phi_\Sigma$$
$$- \partial_k \{G, \underset{b}{v}\}_\Lambda\Big] - \{F, \underset{c}{v}\}_\Lambda{}^\Sigma \Big[\underset{b}{v}{}^m \partial_m \Phi_\Sigma - \{F, \underset{b}{v}\}_\Sigma{}^\Gamma \Phi_\Gamma + \{G, \underset{b}{v}\}_\Sigma\Big]. \tag{3.84}$$

The purpose of the above calculations is that if the commutators of $\underset{a}{\pounds}$ are defined by

$$(\underset{c}{\pounds}, \underset{b}{\pounds})\Phi_\Lambda \equiv (\underset{c}{\pounds}\underset{b}{\pounds} - \underset{b}{\pounds}\underset{c}{\pounds})\Phi_\Lambda, \tag{3.85}$$

then (3.80), (3.81) and (3.84) yield

$$(\underset{c}{\pounds}, \underset{b}{\pounds})\Phi_\Lambda = (\underset{c}{\pounds}\underset{b}{v}{}^k)\partial_k \Phi_\Lambda - \{F, \underset{c}{\pounds}\underset{b}{v}\}_\Lambda{}^\Sigma \Phi_\Sigma - \{G, \underset{c}{\pounds}\underset{b}{v}\}_\Lambda = \underset{cb}{\pounds}\Phi_\Lambda \tag{3.86}$$

where

$$\underset{cb}{v}{}^k = \underset{c}{\pounds}\underset{b}{v}{}^k = -\underset{b}{\pounds}\underset{c}{v}{}^k = \underset{c}{v}{}^m \partial_m \underset{b}{v}{}^k - \underset{b}{v}{}^m \partial_m \underset{c}{v}{}^k. \tag{3.87}$$

Summarizing, we have the following conclusion. Let $\underset{b}{\pounds} f = \underset{b}{v}^m \partial_m f$ *be* $r$ *infinitesimal operators and let* $\Phi_\Lambda$ *be* $N$ *components of a linear geometric object. Then* $(\underset{c}{\pounds}, \underset{b}{\pounds})\Phi_\Lambda$ *is equal to the Lie derivative of* $\Phi_\Lambda$ *with respect to the vector* $\underset{c}{\pounds}\underset{b}{v}^k$ .

If $\underset{a}{\pounds} f = \underset{a}{v}^m \partial_m f$ are $r$ infinitesimal operators of an $r$-parameter continuous group of transformations, we then have

$$(\underset{c}{\pounds}, \underset{b}{\pounds}) f = C_{cb}{}^a \underset{a}{\pounds} f \tag{3.88}$$

where the $C$ is the structure constant of the group and satisfies the relations

$$C_{bc}{}^a = -C_{cb}{}^a , \quad C_{bc}{}^a C_{da}{}^e + C_{cd}{}^a C_{ba}{}^e + C_{db}{}^a C_{ca}{}^e = 0 . \tag{3.89}$$

A trivial calculation then gives

$$\underset{c}{\pounds}\underset{b}{v}^k = C_{cb}{}^a \underset{a}{v}^k \quad [\text{i.e., } (\underset{c}{\pounds}, \underset{b}{\pounds}) f = (\underset{c}{\pounds}\underset{b}{v}^k)\partial_k f] , \tag{3.90}$$

and, hence, we have the following result. If $\underset{a}{\pounds} f$ *are* $r$ *infinitesimal operators of an* $r$*-parameter group of transformations, then we have the formula*

$$(\underset{c}{\pounds}, \underset{b}{\pounds})\Phi_\Lambda = C_{cb}{}^a \underset{a}{\pounds}\Phi_\Lambda \tag{3.91}$$

*for any linear geometric object* $\Phi_\Lambda$. The structure constants of the $r$-parameter group of continuous transformations are thus the same as the structure constants for Lie differentiation of linear geometric objects.

The importance of this result is that $\underset{a}{\pounds} f = \underset{a}{v}^m \partial_m f$ for scalar functions $f$, while the ordinary partial derivative of an object is no longer an object. If we use the Lie derivative, however, the above theorem allows us to extend the commutator relations (3.88) from scalar valued functions to linear geometric objects. When (3.91) is combined with (3.29), we then have

$$'\Phi_\Lambda(x) = (e^{te^a \underset{a}{\pounds}})\Phi_\Lambda \tag{3.92}$$

from the point transformation

$$'x^k = (e^{te^a \underset{a}{\hat{\pounds}}})x^k , \quad \underset{a}{\hat{\pounds}} x^k = \underset{a}{v}^m \partial_m x^k . \tag{3.93}$$

## 3.6. INVARIANCE GROUP OF A GEOMETRIC OBJECT FIELD

Let $\Phi_\Lambda(x^m)$ be a geometric object field of class $p$ on $X_n{}^u$ with $p < u$, and let $G_r$ be an $r$-parameter group of point transformations

$$'x^m = f_m(x^1, \ldots, x^n; \lambda_1, \ldots, \lambda_r) = f_m(x^k; \lambda_a), \quad a = 1, \ldots, r. \quad (3.94)$$

If the group $G_r$ has the property that

$$'\Phi_\Lambda(x) = \Phi_\Lambda(x), \quad (3.95)$$

then we call $G_r$ an *invariance group* of the geometric object field.

Suppose that we can find an infinitesimal point transformation

$$'x^m = x^m + v^m(x^k)\,dt \quad (3.96)$$

on $X_n{}^u$ such that the Lie derivative of a *linear differential* geometric object vanishes:

$$\underset{v}{\pounds}\Phi_\Lambda = v^k \partial_k \Phi_\Lambda - \{F, v\}_\Lambda{}^\Sigma \Phi_\Sigma - \{G, v\}_\Lambda = 0. \quad (3.97)$$

Now, we know that the transformation law of $\Phi_\Lambda$ is given by

$$\Phi_{\Lambda'} = F_\Lambda{}^\Sigma(x^m, x^{m'})\Phi_\Sigma + G_\Lambda(x^m, x^{m'}) \quad (3.98)$$

if $\Phi_\Lambda$ is linear, in which case,

$$\underset{v}{\pounds}\Phi_{\Lambda'} = F_\Lambda{}^\Sigma(x^m, x^{m'})\underset{v}{\pounds}\Phi_\Lambda. \quad (3.99)$$

It thus follows that the equation

$$\underset{v}{\pounds}\Phi_\Lambda = 0 \quad (3.100)$$

has a meaning which does not depend on the choice of the coordinate systems over $X_n{}^u$.

With the vector field $v^k(x^m)$ given, we can choose a coordinate system in a suitable neighborhood of a point of $X_n{}^u$ where $v^k(x^m) \neq 0$ for some value of $k$ such that

$$v^k(x^m) = \delta_1{}^k. \quad (3.101)$$

In this coordinate system, the infinitesimal point transformation

$$'x^m = x^m + \delta_1{}^m\,dt = x^m + v^m(x^k)\,dt \quad (3.102)$$

generates the finite point transformation

$$'x^m = x^m + t\delta_1{}^m . \tag{3.103}$$

From (3.97) and (3.101), we see that for this coordinate system the system of equations (3.100) becomes

$$\underset{v}{\pounds}\Phi_\Lambda = \partial_1 \Phi_\Lambda = 0 , \tag{3.104}$$

and hence the components of the geometric object are independent of the variable $x^1$. Consequently, for the finite point transformation (3.102), we have

$$'\Phi_\Lambda = \Phi_\Lambda[\Phi_\Sigma('x), x^{m'}, x^m] = \Phi_\Lambda \tag{3.105}$$

because $\Phi_\Lambda('x) = \Phi_\Lambda(x)$ and

$$F_\Lambda(\Phi_\Sigma, x^{m'}, x^m) = F_\Lambda(\Phi_\Sigma, x^m, x^m) = \Phi_\Lambda ,$$

as follows from the fact that the object is assumed to be differential. This result can be derived in another way. For the given coordinate system for which (3.101) holds, we have $\underset{v}{\pounds}\Phi = \partial_1 \Phi$, and hence

$$'\Phi_\Lambda = (e^{t\underset{v}{\pounds}})\Phi_\Lambda = \Phi_\Lambda + t\underset{v}{\pounds}\Phi_\Lambda + t^2 \underset{v}{\pounds}\underset{v}{\pounds}\Phi_\Lambda + \cdots$$

From these equations we see that (3.100) implies (3.105)

**Theorem 3.1.** *If a space admits an infinitesimal point transformation with respect to which the Lie derivative of a linear differential geometric object vanishes, then it admits a one-parameter invariance group of this geometric object.* The importance of Theorem 3.1 is that it allows us to work solely with the infinitesimal transformations of a group when dealing with linear differential geometric objects. It says in effect that if we know what happens for a sufficiently small neighborhood of the identity transformation of the one-parameter group (that is, $\underset{v}{\pounds}\Phi_\Lambda = 0$), then we can extend that knowledge to the entire one-parameter group.

**Theorem 3.2.** *In order that a space admit a one-parameter invariance group of a linear differential geometric object, it is both necessary and sufficient that there exist a coordinate system with respect to which the components of the geometric object are independent of one of the coordinates.*

Suppose that $r$ contravariant vectors $v_a{}^k(x^m)$ define $r$ one-parameter invariance groups of the same linear differential geometric object $\Phi_\Lambda$, then we have

$$\underset{a}{\pounds}\Phi_\Lambda = \underset{a}{v}^m \partial_m \Phi_\Lambda - \{F, v_a\}_\Lambda{}^\Sigma \Phi_\Sigma - \{G, v_a\}_\Lambda = 0 \tag{3.106}$$

and hence

$$C^a \underset{a}{\pounds}\Phi_\Lambda = C^a v_a{}^m \partial_m \Phi_\Lambda - \{F, C^a v_a\}_\Lambda{}^\Sigma \Phi_\Sigma - \{G, C^a v_a\}_\Lambda = 0 \tag{3.107}$$

in which $C^a$ is a constant.

**Theorem 3.3.** *If each of $r$ contravariant vectors generates a one-parameter invariance group of a given linear differential geometric object, then a linear combination of these contravariant vectors with constant coefficients also generates a one-parameter invariance group of the same geometric object.*

**Theorem 3.4.** *If each of $r$ vectors $v_a{}^k(x^m)$ defines a one-parameter invariance group of a given linear differential geometric object, then each of the vectors $\underset{a}{\pounds}\underset{b}{v}^k(x^m)$ defines a one-parameter invariance group of this same geometric object.* A straightforward combination of the above result and the italicized statement after equation (3.87) gives the following result.

**Theorem 3.5.** *If each of $r$ infinitesimal operators of an $r$-parameter group of transformations generates a one-parameter invariance group of a given linear differential geometric object, then all transformations of the $r$-parameter group leave invariant the geometric object; that is, the group is an invariance group.*

Suppose that each of $r$ linearly independent vectors $v_a{}^m(x^k)$ defines a one-parameter invariance group of some linear differential geometric object. The vectors $v_a{}^k(x^m)$ are linearly independent if $C^a v_a{}^k(x^m) = 0$, with $C^a =$ constants, if and only if $C^a = 0$. If, for any vector $v^k(x^m)$ which defines a one-parameter invariance group of this geometric object, we have $v^k(x^m) = C^a v_a{}^k(x^m)$, then the set of vectors $v_a{}^k(x^m)$ is said to be *complete*. Now, if $r$ vectors $v_a{}^k(x^m)$ form a complete set of vectors defining $r$ one-parameter invariance groups of a given linear differential object, then, since $\underset{a}{\pounds}\underset{b}{v}^k(x^m)$ are also vectors defining an invariance group, we must have

$$\underset{a}{\pounds}\underset{b}{v}^k(x^m) = C_{ab}{}^c v_c{}^k(x^m) \tag{3.108}$$

in which $C_{ab}{}^{c}$ are constants. The equations (3.108) show that $\underset{a}{£} f = \underset{a}{v}{}^{k} \partial_k f$ are $r$ infinitesimal operators of an $r$-parameter group.

**Theorem 3.6.** *If $r$ vectors $\underset{a}{v}{}^{k}(x^m)$ form a complete set of vectors defining $r$ one-parameter invariance groups of a given linear differential geometric object, then $\underset{a}{£} f = \underset{a}{v}{}^{k} \partial_k f$ are $r$ infinitesimal operators of an r-parameter invariance group of the object.* Hence for any constants $C^a$,

$$'x^m = (e^{tC^a \underset{a}{£}})\, x^m$$

defines a finite invariance transformation.

The theorems of Lie are central to the considerations of Sec. 3.6, and thus are stated for the convenience of the reader.

I.1. If the invertible transformations

$$'x^m = f^m(x^k;\, \lambda^a)\,, \quad a = 1, \ldots, r \tag{3.109}$$

form an $r$-parameter group, then there exist $r^2$ functions $A_a{}^b(\lambda^c)$ such that $\det(A_a{}^b) \neq 0$ and $nr$ functions $v_b{}^m(x^k)$ for which no equation exists of the form $C^a v_a{}^m = 0$ with coeffieients $C^a$ independent of $x^k$, such that

$$\frac{\partial' x^m}{\partial \lambda^a} = A_a{}^b(\lambda^c)\, v_b{}^m('x^k)\,. \tag{3.110}$$

I.2. If a set of transformations,

$$'x^m = f^m(x^k;\, \lambda^a)\,, \quad a = 1, \ldots, r\,, \tag{3.111}$$

is given that contains the identity transformation and is such that $\det(\partial' x^m/\partial x^k) \neq 0$ in some neighborhood of the identity transformation, if there exist $r^2$ function $A_a{}^b(\lambda^c)$ such that $\det(A_a{}^b) \neq 0$, and if there exist $nr$ functions $v_b{}^m(x^k)$ that do not satisfy any equation of the form $C^a v_a{}^m = 0$ with $C$ independent of $x$ and such that (3.110) holds, then the transformations (3.111) form an $r$-parameter group.

II.1 If

$$X_b \equiv v_b{}^m(x^k)\frac{\partial}{\partial x^m}\,, \quad b = 1, \ldots, r\,, \tag{3.112}$$

are the symbols of $r$ independent (with constant coefficients) infinitesimal transformations of $x$ that satisfy the equations

$$(X_a, X_b)f = C_{ab}{}^{c} X_c f \tag{3.113}$$

for all $C^2$ functions $f(x^k)$ and if the coefficients $C_{ab}{}^{c}$ are constants and such that

$$C_{ab}{}^{c} + C_{ba}{}^{c} = 0\,, \tag{3.114}$$

then the $r$ infinitesimal transformations with symbols $X_b$ are the infinitesimal transformations of an $r$-parameter group.

II.2. If $X_a$ are the symbols of $r$ independent infinitesimal transformations of an $r$-parameter group, then equations (3.113) hold, in which $C$ are constants and satisfies (3.114).

III.1. If $X_a$ is the symbol of $r$ independent infinitesimal transformations of an $r$-parameter group, then the $C$ that appears in equations (3.113) satisfies the relations

$$C_{ab}{}^{e}C_{ce}{}^{d} + C_{bc}{}^{e}C_{ae}{}^{d} + C_{ca}{}^{e}C_{be}{}^{d} = 0\,. \tag{3.115}$$

III.2. If $\frac{1}{2}r^2(r-1)$ constants $C_{ab}{}^{c}$ satisfy equations (3.114) and (3.115), then there always exists an $r$-parameter group with $r$ independent infinitesimal transformation symbols $X_b$ such that $X_b$ satisfy the equations (3.113).

## Problems

3.1 Let $x^{m'} = f_m(x^1, \ldots, x^n) = f_m(x^k)$ be an element of $G_n{}^u$ and set $\Delta = \det(\partial x^{m'}/\partial x^k) = \det(\partial f_m/\partial x^k)$. Find the transformation function $F_\Lambda(\Phi_\Sigma, x^m, x^{m'})$ for each of the following cases and show that these transformation functions possess the group property.

a. A scalar function: $N = 1$

$$\Phi'(x^{m'}) = \Phi[x^k(x^{m'})]\,.$$

b. A contravariant vector: $N = n$

$$v^{k'}(x^{m'}) = \frac{\partial x^{k'}}{\partial x^m}\, v^m[x^r(x^{t'})]\,.$$

c. A covariant vector: $N = n$

$$\omega_{k'}(x^{m'}) = \frac{\partial x^m}{\partial x^{k'}}\,\omega_m[x^r(x^{t'})]\,.$$

d. A tensor of contravariant valence $p$ and covariant valence $q$; $N = n^{p+q}$

$$P^{k'_1 \ldots k'_p}_{m'_1 \ldots m'_q}(x^{t'}) = \frac{\partial x^{k'_1}}{\partial x^{s_1}} \cdots \frac{\partial x^{k'_p}}{\partial x^{s_p}} \frac{\partial x^{t_1}}{\partial x^{m'_1}} \cdots \frac{\partial x^{t_q}}{\partial x^{m'_q}} P^{s_1 \ldots s_p}_{t_1 \ldots t_q}[x^u(x^{v'})] .$$

The class of all such collections of functions is said to form the *tensor class* $T(p,q)$. The class of all scalars is $T(0,0)$, the class of all contravariant vectors is $T(1,0)$, and the class of all covariant vectors is $T(0,1)$.

e. If a collection of functions transforms as a member of $T(p,q)$, but with a multiplicative factor $\tau$ dependent on $\Delta$, then it is a
   1) Tensor $\Delta$-density of weight $\omega$ and antiweight $\omega'$ if $\tau = \Delta^{-\omega}\bar{\Delta}^{-\omega'}$ with $\bar{\Delta}$ = complex conjugate of $\Delta$;
   2) Tensor density of weight $\omega$ if $\tau = |\Delta|^{-\omega}$;
   3) $W$-tensor if $\tau = \Delta/|\Delta|$.

Note that a $W$-scalar with the value 1 in some coordinate system distinguishes transformations with $\Delta > 0$ from those with $\Delta < 0$.

f. A linear connection, $N = n^3$,

$$L^{k'}_{\ell' m'}(x^{t'}) = \frac{\partial x^s}{\partial x^{\ell'}} \frac{\partial x^t}{\partial x^{m'}} \left\{ \frac{\partial x^{k'}}{\partial x^u} L_{st}{}^u[x^v(x^{\omega'})] - \frac{\partial}{\partial x^s} \frac{\partial x^{k'}}{\partial x^t} \right\} .$$

It is useful to introduce

$$A_t{}^{k'} = \frac{\partial x^{k'}}{\partial x^t} \qquad A_{k'}{}^t = \frac{\partial x^t}{\partial x^{k'}}$$

so that

$$A_t{}^{k'} A_{s'}{}^t = \delta_{s'}{}^{k'} , \quad A_t{}^{k'} A_{k'}{}^m = \delta_t{}^m .$$

We then have

$$L^{k'}_{\ell' m'} = A_{\ell'}{}^s A_{m'}{}^t (A_u{}^{k'} L_{st}{}^u - \partial_s A_t{}^{k'}) .$$

Examples a - e are linear, homogeneous, differential geometric objects and example f is a linear, differential (nonhomogeneous) geometric object.

3.2 Let $R_{km} \in T(0,2)$ such that (i) $\det(R_{km}) \neq 0$, (ii) $R_{km} = R_{mk}$, and let $X_n{}^u$ be such that there exists a coordinate system $(x)$ on

$X_n{}^u$ such that $R_{km}$ is diagonal and $R_{km} = \text{diag}(+1, +1, +1, -1)$. Let $A_{km} \in T(0,2)$ such that (i) $A_{km} = -A_{mk}$, (ii) $a = \det(A_{km}) \neq 0$, and (iii) $4A = A_{km}A_{st}R^{ks}R^{mt} \neq 0$ where $R^{ks}R_{st} = \delta_t^k$. Finally, let $n = 4$ ($n$ obviously must be even if $a \neq 0$). A vector $U^k \in T(1,0)$ is said to be *an eigenvector of* $A_{km}$ *relative to* $R_{km}$ *with eigenvalue* $\rho$ *iff*

$$(A_{km} - \rho R_{km})U^m = 0 .$$

It is easily shown that there are only four eigenvectors $U_1{}^m, U_2{}^m, U_3{}^m, U_4{}^m$ (to within factors of proportionality) and that

a. $U_1{}^k, U_2{}^k$, are complex conjugates, $U_3{}^k, U_4{}^k$ are real;
b. $R_{km}U_1{}^kU_2{}^m = R_{km}U_3{}^kU_4{}^m = 1$ and $R_{km}U_a{}^kU_b{}^m = 0$ for all other independent combinations of $a$ and $b$:
c. The eigenvectors are unique to within the transformations

$$(*)\begin{cases} 'U_1{}^k = e^{i\eta}U_1{}^k , & 'U_2{}^k = e^{-i\eta}U_2{}^k \\ 'U_3{}^k = \epsilon e^{\mu}U_3{}^k , & 'U_4{}^k = \epsilon e^{-\mu}U_4{}^k \\ \epsilon = \pm 1 , & \eta \text{ and } \mu \text{ real valued.} \end{cases}$$

Define the matrices $\mathbf{G}_k$ and $\mathbf{N}^k$ by

$$\mathbf{G}_k = \begin{pmatrix} 0 & 0 & U_k{}^1 & U_k{}^3 \\ 0 & 0 & U_k{}^4 & -U_k{}^2 \\ U_k{}^2 & U_k{}^3 & 0 & 0 \\ U_k{}^4 & -U_k{}^1 & 0 & 0 \end{pmatrix}$$

$$\mathbf{N}^k = \begin{pmatrix} 0 & 0 & U_1{}^k & U_4{}^k \\ 0 & 0 & U_3{}^k & -U_1{}^k \\ U_1{}^k & U_4{}^k & 0 & 0 \\ U_3{}^k & -U_2{}^k & 0 & 0 \end{pmatrix}$$

where $U_k{}^a$ is uniquely defined by

$$U_k{}^a U_b{}^k = \delta_b{}^a \quad \text{or} \quad U_k{}^a U_a{}^m = \delta_k{}^m \,.$$

It is then easily shown that

$$\mathbf{G}_k \mathbf{N}^k = \mathbf{N}^k \mathbf{G}_k = 2\mathbf{E} \,, \qquad \mathbf{E} = \mathrm{diag}(1,1,1,1) \,,$$

$$\mathbf{G}_k \mathbf{G}_m + \mathbf{G}_m \mathbf{G}_k = R_{km} \mathbf{E} \,,$$

$$\mathbf{N}^k \mathbf{N}^m + \mathbf{N}^m \mathbf{N}^k = R^{km} \mathbf{E} \,.$$

With $v^k \in T(1,0)$, we associate

$$\mathbf{v} = \mathbf{G}_k v^k$$

and with $\omega_k \in T(0,1)$, we associate

$$\boldsymbol{\omega} = \mathbf{N}^k \omega_k \,.$$

It is then easily shown that

$$v^k \mathbf{E} = (\mathbf{v}\mathbf{G}_m + \mathbf{G}_m \mathbf{v}) R^{mk}$$

and

$$\omega_k \mathbf{E} = (\boldsymbol{\omega}\mathbf{N}^m + \mathbf{N}^m \boldsymbol{\omega}) R_{mk}$$

so that $v^k$ and $\omega_k$ can be recovered from $\mathbf{v}$ and $\boldsymbol{\omega}$, respectively. When the eigenvectors $U_a{}^k$ undergo the transformations (*) we have

$$'\mathbf{G}_k = \mathbf{C}^{-1} \mathbf{G}_k \mathbf{C} \,, \qquad '\mathbf{N}^k = \mathbf{C} \mathbf{N}^k \mathbf{C}^{-1}$$

with

$$\mathbf{C} = \mathbf{C}(\Phi) = \begin{pmatrix} \ell_1 e^{-\Phi} & 0 & 0 & 0 \\ 0 & \ell_1 e^{\Phi} & 0 & 0 \\ 0 & 0 & \ell_2 e^{-\bar{\Phi}} & 0 \\ 0 & 0 & 0 & \ell_2 e^{\bar{\Phi}} \end{pmatrix}$$

$$C^{-1} = C(-\Phi)\,, \quad \ell_1 = \pm 1\,, \quad \ell_2 = \pm 1\,, \quad 2\Phi = \eta + i\left(\mu + (1-\epsilon)\frac{\pi}{2}\right).$$

Accordingly,

$$'\mathrm{v} = C^{-1}\mathrm{v}C\,, \quad '\boldsymbol{\omega} = C\boldsymbol{\omega}C^{-1}\,.$$

Objects with the transformation laws are referred to as analytic spinors; v is of type $(\bar{1}, 1)$ and $\boldsymbol{\omega}$ is of type $(1, \bar{1})$.

Now, it is trivial to show that v and $\boldsymbol{\omega}$ transform as matrices of scalars under transformations of coordinates as applied to the vectors $U^k$, $\omega_k$, $U_a{}^k$, and $U_k{}^a$. On the other hand, suppose that the real valued quantities $\eta$ and $\mu$ are functions of the old and new coordinates. What properties must these functions have so that analytic spinors of type $(1, \bar{1})$ and of type $(\bar{1}, 1)$ are components of geometric objects? In answering this question, you should examine whether or not any elements of $G_n{}^u$ can be used or must the transformations be restricted to a sub-pseudo group, in particular, to the general linear group or the group which leaves $R_{ap}$ invariant? Note, also, that the vectors $U_a{}^m$ should be considered as collections of point-valued functions while $\eta(x^m)$, $\mu(x^m)$ correlate the $U_a{}^m$ at different points.

3.3 Show that

a. $\underset{v}{£}\Phi = v^\mu \partial_\mu \Phi \quad \text{for} \quad \Phi \in T(0,0)\,,$

$\underset{v}{£}\Phi = v^\mu \partial_\mu \Phi + \omega\Phi\partial_\mu v^\mu$ for $\Phi$ a scalar density of weight $\omega$.

b. $\underset{v}{£}\Phi_m = v^k\partial_k\Phi_m + \Phi_k\partial_m v^k \quad \text{for} \quad \{\Phi_m\} \in T(0,1)\,,$

$\underset{v}{£}\Phi_m = v^k\partial_k\Phi_m + \Phi_k\partial_m v^k + \omega\Phi_m\partial_k v^k$

for $\{\Phi_m\}$ a covariant vector density of weight $\omega$.

c. $\underset{v}{£}\Phi^m = v^k\partial_k\Phi^m - \Phi^k\partial_k v^m \quad \text{for} \quad \{\Phi^m\} \in T(1,0)\,,$

$\underset{v}{£}\Phi^m = v^k\partial_k\Phi^m - \Phi^k\partial_k v^m + \omega\Phi^m\partial_k v^k$

for a contravariant vector density of weight $\omega$.

d. $\underset{v}{£}\Phi^{m_1\cdots}_{k_1\cdots} = v^t\partial_t\Phi^{m_1\cdots}_{k_1\cdots} + \Phi^{m_1\cdots}_{t\cdots}\partial_{k_1}v^t + \cdots$

$- \Phi_{k_1}{}^t\partial_t v^{m_1} - \cdots + \omega\Phi^{m_1\cdots}_{k_1\cdots}\partial_t v^t$

for $\{\Phi^{m_1 \cdots}_{k_1 \cdots}\} \in T(p,q)$ of weight $\omega$. Hint: use the fact that $T(p,q)$ can be obtained from the Cartesian product

$$\underbrace{T(1,0) x \cdots x T(1,0)}_{p} \underbrace{x T(0,1) x \cdots x T(0,1)}_{q}$$

and that the result is linear, homogeneous, and differential so that $\underset{v}{£}$ satisfies the rule of Leibniz.

3.4 Let $\{\Phi_\Lambda\}$ be a geometric object field in $X_n{}^u$ such that there exists a coordinate system of $X_n{}^u$ for which each components of $\{\Phi_\Lambda\}$ has a constant value on $X_n{}^u$. Show that

$$\underset{a}{v}{}^m(x) = \delta_a{}^m, \quad a = 1, \ldots, m$$

are the infinitesimal generators of an invariance group for $\{\Phi_\Lambda\}$.

3.5 If $n = 3$ and $g_{km} = \text{diag}(1,1,1)$ in some coordinate system over $X_3{}^u$ with $\{g_{km}\} \in T(0,2)$, then

$$\underset{1}{v}{}^m = \delta_1{}^m, \quad \underset{2}{v}{}^m = \delta_2{}^m, \quad \underset{3}{v}{}^m = \delta_3{}^m, \quad \underset{4}{v}{}^m = x^1\delta_2{}^m - x^2\delta_1{}^m,$$

$$\underset{5}{v}{}^m = x^2\delta_3{}^m - x^3\delta_2{}^m, \quad \underset{6}{v}{}^m = x^3\delta_1{}^m - x^1\delta_3{}^m$$

is an invariance group for $\{g_{km}\}$.

3.6 Suppose that $\underset{v}{£} g_{km} = \Phi g_{km}$ for $\{g_{km}\} \in T(0,2)$, $\Phi \in T(0,0)$ and suppose that $g_{km} = g_{mk}$, $\det(g_{km}) \neq 0$. Show that $v^k$ generates a one-parameter invariance group for the geometric object field

$$G_{km} = g_{km}\{|\det(g_{st})|\}^{1/n}.$$

The vector field $v^k$ is said to determine a *conformal motion of* $X_n{}^u$ when $|g_{km} P^k P^m|^{1/2}$ is used to define a pseudonorm on the elements of $T(1,0)$ over $X_n{}^u$. This pseudonorm is a norm iff $g_{km}$ is positive definite.

3.7 Let $n = 4$ and $g_{km} = \text{diag}(1,1,1,-1)$ define a pseudonorm on $X_4{}^u$ for some coordinate covering of $X_4{}^u$. Show that the full invariance group of $\{g_{km}\}$ is the component of the Lorentz group that is continuously connected to the identity element. An $r$-parameter group for which

$$\underset{a}{\pounds} g_{km} = 0, \quad a = 1, \ldots, r$$

is said to define a group of "metric motions" of $X_n^u$ if $\{g_{km}\}$ is symmetric, nonsingular, and defines a pseudonorm on $T(1,0)$ over $X_n^u$. If $g_{km} = \text{diag}(1,1,1,-1)$, $X_4^u$ is referred to as Minkowsky space.

Conclude that the component of the Lorentz group that is continuously connected to the identity element is a complete group of metric motions in Minkowsky space.

3.8 Define $u^k(x^m)$ by $u^k(x^m) = C^a \underset{a}{v}{}^k(x^m)$, where $C^a$ are $r$ constants. Show that

$$\underset{u}{\pounds} \Phi_\Lambda = C^a \underset{v\,a}{\pounds} \Phi_\Lambda$$

for any linear geometric object $\Phi_\Lambda$. Investigate the form of $\underset{u}{\pounds} \Phi_\Lambda$ for a general (nonlinear) geometric object.

3.9 Prove that

$$\underset{t}{\pounds}\underset{u}{\pounds} v^k + \underset{u}{\pounds}\underset{v}{\pounds} t^k + \underset{v}{\pounds}\underset{t}{\pounds} u^k = 0$$

on any domain where the vectors $t^k, u^k, v^k \in T(1,0)$ are of class $C^2$.

3.10 Let $u^k(x^m)$ and $v^k(x^m)$ be analytic vector fields on a domain $D^*$ of $X_n$. Show that

$$e^{-s\underset{u}{\pounds}} e^{-t\underset{v}{\pounds}} e^{+s\underset{u}{\pounds}} \Phi_\Lambda = e^{-t\underset{w}{\pounds}} \Phi_\Lambda, \quad w^k = e^{-s\underset{u}{\pounds}} v^k$$

for any linear geometric object $\Phi_\Lambda$.

# 4 Invariance Considerations

## 4.1. PRELIMINARY CONSIDERATIONS

Repeated use is made of the Emmy Noether theorems in many contexts. These theorems relate conservation laws satisfied by solutions of Euler equations and the invariance of functionals (which gives rise to the Euler equations) when the arguments of these functionals are changed by a coordinate transformation or a point transformation acting on the manifold of independent variables. Although the orignal forms of the Noether theorems are of use in many problems, significant extensions of results can be obtained if we approach the problem from the standpoint that the functions occurring as arguments of the functionals under consideration are geometric objects. From a physical point of view, it would appear almost necessary to assume that the functions which appear in the lagrangian function are geometric objects; otherwise, there would be no direct method in which one-to-one correlations could be obtained between different observers and different frames of reference.

The geometric objects of class $p$ considered in Chap. 3 are such that we can associate components of such objects with every admissible coordinate transformation of the pseudogroup $G_n{}^u$. Our considerations are now restricted to classes of objects that have well defined transformation properties under certain subgroups or subpseudogroups of $G_n{}^u$. The reason for this restriction is that we are often only able to establish that a specific group $G$ of transformation induces invariance properties. This is particularly the case in physical problems in which the Galilean group or the Lorentz group plays a central role.

If an ordered collection of $N$ functions $\Phi_\Lambda(x^m)$ transforms according to the law of a geometric object of class $p$ under coordinate transformations of a given subgroup or subpseudogroup $G$ of the admissible pseudogroup $G_n^u$, then $\{\Phi_\Lambda(x^m)\}$ is said to form a *geometric object of class* $p$ *relative to* $G$.

The collection of all geometric objects of class $p$ relative to $G$, with $N$ components and with fixed functional forms of $F_\Lambda$ that define the transformation laws, is said to form a *geometry class* $\mathcal{G}(p, N; F_\Lambda)/G$ *relative to* $G$. If $G$ is not a proper subgroup or subpseudogroup of $G_n^u$, the $\mathcal{G}(p, N; F_\Lambda)/G$ will be written as $\mathcal{G}(p, N; F_\Lambda)$ and will be referred to simply as a *geometry class*.

It is evident from the above definitions that if $G(1)$ and $G(2)$ are subgroups of the admissible group, then $\mathcal{G}(p, N; F_\Lambda)/G(1)$ is contained in $\mathcal{G}(p, N; F_\Lambda)/G(2)$ accordingly as $G(2)$ is contained in $G(1)$. For example, the geometry class $(G)$ relative to the general linear group consisting of all tensors of the form $T_{mn}^{\ \ k}$ contains the components of affine connection of the space as an element. For any group which consists of other than linear transformations, the components of affine connection cease to be elements of $(G)$ (recall that the functional form of $F_\Lambda$ remains fixed).

From the definition of $\mathcal{G}(p, N; F_\Lambda)/G$, it easily follows that if the components $\Phi_\Lambda(x)$ of an element of $\mathcal{G}(p, N; F_\Lambda)/G$ are $C^v$ functions on $D^*$, then $\Phi_{\Lambda'}(x)$ are $C^w$ functions on $D^*$, where $w = \min(u - p, v)$. We shall assume throughout Chap. 4 that $u > p + 2$. This will ensure that the collection of all elements of $\mathcal{G}(p, N; F_\Lambda)/G$, whose components are of class $C^r$ with $r \leq 2$ in any one coordinate system, will not have the continuity class of its components degraded by the action of $G$. Under this assumption it is obvious that $\mathcal{G}(p, N; F_\Lambda)/G \subset \mathfrak{D}_1(D^*; N)$, and hence we may use elements of a geometry class as arguments of a lagrangian function.

In practice, it is usual to take the underlying group $G_n^u$ as the group $G_A$ of all analytic coordinate transformations on $D$. In this case, we can actually identify $\mathcal{G}(p, N; F_\Lambda)/G_A$ with $\mathfrak{D}_1(D^*; N)$, since any subgroup of $G_A$ induces a map of $\mathfrak{D}_1(D^*; N)$ onto $\mathfrak{D}_1(D^*; N)$.

If $\{\Phi_\Lambda(x^m)\} \in \mathcal{G}(p, N; F_\Lambda)$, then the collection of $N(n+1)$ functions $\{\Phi_\Lambda, \partial_m \Phi_\Lambda\}$ are also the components of a geometric object (see (3.17)), which we denote by $\{_\alpha\Phi_\Lambda(x^m)\}$, $\alpha = 0, 1, \ldots, n$, where

$$_0\Phi_\Lambda(x^m) \equiv \Phi_\Lambda(x^m), \quad {}_m\Phi_\Lambda(x^m) \equiv \partial_m \Phi_\Lambda(x^m). \tag{4.1}$$

For convenience, we denote the transformation functions of $_\alpha\Phi_\Lambda$ by $_\alpha F_\Lambda({}_\beta\Phi_\Sigma, x^m, x^{m'})$. By the *derived geometry class*, $\partial\mathcal{G}(p, N; F_\Lambda)/G$, relative to G, we mean the collection of all elements $\{_\alpha\Phi_\Lambda(x^m)\}$ for all $\{\Phi_\Lambda(x^m)\} \in \mathcal{G}(p, N; F_\Lambda)/G$ in which the ordering is given by (4.1).

Under the previous continuity assumptions, we have that $\partial \mathcal{G}(p, N; F_\Lambda)/G \subset \mathfrak{D}_1(D^*; N(n+1))$.

Whenever we write $\{\Phi_\Lambda(x^m)\}$ throughout the remainer of Chap. 4, *we shall assume*, unless stated to the contrary, *that* $\{\Phi_\Lambda(x^m)\}$ *belongs to a geometry class relative to some group* $G$. Trivially, there is no loss of generality, for if $G$ is the group I consisting of only the identity element, then every element of $\mathfrak{D}_1(D^*; N)$ is an element of this geometry class. The reason for not specifying $G$ *a priori* is that we often seek equations for the determination of $G$ such that certain specific conditions will be satisfied.

Let $g_a(x^m, z^m, {}_\alpha\Phi_\Lambda)$ be $q$ arbitrary but fixed functions of class $C^2$ in their $N(n+1) + 2n$ arguments for all $x$ and $z$ in $D^*$ and all $\Phi_\Lambda(x^m)$ in $\mathfrak{D}_1(D^*; N)$. It should be carefully noted that no particular transformation properties are assumed for the functions $g_a$ when its arguments are transformed by the action of elements of $G_n{}^u$. If $d$ is any compact, simply connected, $n$-dimensional set contained in $D^*$, we define functionals $k_a(x^m, \{\Phi_\Lambda\}; d)$ by the relations

$$k_a(x^m, \{\Phi_\Lambda\}; d) = \int_d g_a[x^m, z^m, {}_\alpha\Phi_\Lambda(z^m)]\, dV(z) \,. \tag{4.2}$$

It is obvious that $k_a(x^m, \{\Phi_\Lambda\}; D^*) = k_a(x^m, \{\Phi_\Lambda\})$, where the functionals $k_a$ appearing on the right-hand side of the equality are the same $k_a$ considered in Chaps. 1 and 2. Similarly, if

$$L = L[x^m, {}_\alpha\Phi_\Lambda(x^m), k_a(x^m, \{\Phi_\Lambda\}; d)] \,, \tag{4.3}$$

we define the functionals $J[\{\Phi_\Lambda\}](L; d)$ by the relations

$$J[\{\Phi_\Lambda\}](L; d) = \int_d L[x^m, {}_\alpha\Phi_\Lambda(x^m), k_a(x^m, \{\Phi_\Lambda\}; d)]\, dV(x) \tag{4.4}$$

in which $J[\{\Phi_\Lambda\}](L; D^*) = J[\{\Phi_\Lambda\}](L)$.

Since the functions $\{\Phi_\Lambda(x^m)\}$ appearing in (4.4) are arbitrary elements of $\mathfrak{D}_1(D^*; N)$, it is meanginful to ask whether there exists a particular set $S(L)$ of $\mathfrak{D}_1(D^*; N)$ which has the property of rendering the functional $J[\{\Phi_\Lambda\}](L; D^*)$ stationary in value under the assumption that $L$ is a fixed function of its arguments and that $\{\Phi_\Lambda\}$ assume given values on $\partial D^*$. This is the question examined in Chap. 2, and hence we know the answer: the set $S(L)$ is comprised solely of solutions of the Euler equations

$$\{\mathcal{E}|L\}_{\Phi_\Lambda}(x^m) = 0 \,. \tag{4.5}$$

However, a slightly different notation for the variation process is introduced since our interests in this chapter lie along different lines than those of Chap. 2.

Let $\Phi_\Lambda(x^m; \eta)$ be a one-parameter family of elements of $\mathfrak{D}_1(D^*; N)$ with the following properties: (i) they are analytic in $\eta$ for all $\eta$ in some interval $(a, b)$ of the real line that contains the identity element of addition (the zero element), (ii) $\Phi_\Lambda(x^m; 0) = \Phi_\Lambda(x^m)$, and (iii) $\Phi_\Lambda(x^m; \eta)|_{\partial D} = \Phi_\Lambda(x^m)|_{\partial D}$.

We then write

$$\Phi_\Lambda(x^m; \eta) = \Phi_\Lambda(x^m) + \eta\delta\Phi_\Lambda(x^m) + o(\eta) \tag{4.6}$$

where

$$\delta\Phi_\Lambda(x^m) = \left.\frac{\partial\Phi_\Lambda(x^m; \eta)}{\partial\eta}\right|_{\eta=0} \subset \mathfrak{D}_1(D^*, N) \tag{4.7}$$

and

$$\delta\Phi_\Lambda(x^m)|_{\partial D} = 0 . \tag{4.8}$$

Accordingly, by the considerations of Chaps. 1 and 2, we obtain

$$\Delta J[\{\Phi_\Lambda\}](L; D) = \eta \int \{\mathcal{E}|L\}_{\Phi_\Lambda}(x^m)\,\delta\Phi_\Lambda(x^m)\,dV(x) + o(\eta) . \tag{4.9}$$

This formulation obviously extends to the case of a vector $\{\eta^t\}$, $t = 1, \ldots, T$ in $T$-dimensional euclidean space $E_t$. If $\|\eta\|$ denotes the norm of $\{\eta^t\}$ in $E_t$, then

$$\left\{\begin{aligned} \Phi_\Lambda(x^m; \eta^t) &= \Phi_\Lambda(x^m) + \eta^t\delta_t\Phi_\Lambda(x^m) + o(\|\eta\|) , \\ \delta_s\Phi_\Lambda &= \left.\frac{\partial\Phi_\Lambda(x^m; \eta^t)}{\partial\eta^s}\right|_{\|\eta\|=0} \in \mathfrak{D}_1(D^*; N) , \\ \delta_s\Phi_\Lambda(x^m)|_{\partial D} &= 0 . \end{aligned}\right. \tag{4.10}$$

## 4.2. COORDINATE TRANSFORMATIONS AND EXTENDED VARIATIONS

Consider the one-parameter family $T(\lambda)$ of coordinate transformations

$$x^{m'} = G^m(x^k, \lambda) \tag{4.11}$$

of class $C^u$ in $x^k$ for all values of the parameter $\lambda$ in some interval $(k, m)$ of the real line containing the identity of addition, and such that

$$G^m(x^k, \lambda) = x^m + \lambda v^m(x^k) + o(\lambda), \quad \det\left(\frac{\partial G^m}{\partial x^k}\right) \neq 0, \tag{4.12}$$

where

$$v^m(x^k) = \left.\frac{\partial G^m(x^k, \lambda)}{\partial \lambda}\right|_{\lambda=0} \in C^u \tag{4.13}$$

is referred to as the *infinitesimal generator* of the transformations $T(\lambda)$.

In a sufficiently small neighborhood of $\lambda = 0$, the equations of transformation for the element of $\mathcal{G}(p, N; F_\Lambda)$ give

$$\Phi_{\Lambda'}(x) = \Phi_\Lambda(x) + \lambda \underset{v}{D} \Phi_\Lambda(x) + o(\lambda), \tag{4.14}$$

where

$$\underset{v}{D} \Phi_\Lambda(x) = v^k \partial_k \Phi_\Lambda - \underset{v}{\pounds} \Phi_\Lambda \tag{4.15}$$

as dervied in Chap. 3 (equations (3.26) and the realted material). In view of (4.14), we refer to the quantity $\underset{v}{D}\Phi_\Lambda$ as the coordinate veration of $\Phi_\Lambda$ which is induced by the transformation $T(d\lambda)$. The quantity $\underset{v}{D}\Phi_\Lambda$ has been shown also to be the increment in $\Phi_\Lambda$ which results from the dragging along of $\Phi_\Lambda(x^m)$ over $v^m(x^k)\,d\lambda$ by the transformation $T(\lambda)$. The reader should recall that this was the second method of development in which the considerations were based on $T(\lambda)$ rather than on $T(\lambda)^{-1}$. Specifically, recall that the coordinate system dragged along by $T^{-1}$ was given by $x^{m'} = G^m(x^k, \lambda)$ when the points were subjected to the point transformation $'x^m = G^m(x^k, \lambda)$. The corresponding relations based on $T(\lambda)$ read

$$'x^m = G^m(x^k, \lambda), \quad x^{m'} = g^m(x^k, \lambda), \quad G \circ g = 1. \tag{4.16}$$

It thus follows that *a coordinate transformation with infinitesimal generator* $v^m(x^k)$ *induces a change in the field* $\{\Phi_\Lambda(x^m)\}$ *which is exactly the change which would result from dragging along* $\{\Phi_\Lambda(x^m)\}$

*over* $v^m(x^k)d\lambda$ *by* $T(d\lambda)$. This observation will be of repeated use later.

If we compose coordinate variation with function variation in $\mathfrak{D}_1(D^*; N)$, we then have *the extended variation*

$$\begin{aligned}\Phi_{\Lambda'}(x^m, \lambda, \eta) &\equiv \tilde{\Phi}_{\Lambda'}(x) \\ &= \Phi_\Lambda(x) + \lambda \underset{v}{D}\Phi_\Lambda(x) + \eta\delta\Phi_\Lambda(x) + o(\lambda, \eta) .\end{aligned} \tag{4.17}$$

The collection of functions

$$\underset{v}{\Delta}\Phi_\Lambda(x) = \underset{v}{D}\Phi_\Lambda(x) + \delta\Phi_\Lambda(x) \tag{4.18}$$

is referred to as the *total infinitesimal variation* of $\Phi_\Lambda(x)$ which results under the confluence $\lambda = \pi = \eta$. If the total variation vanishes, we then have

$$\delta\Phi_\Lambda(x) = -\underset{v}{D}\Phi_\Lambda(x) ;$$

and thus, $\delta\Phi_\Lambda(x)$ is the change in $\Phi_\Lambda$ that results from dragging along $\Phi_\Lambda$ over $-v^m(x^k)d\lambda$ by $T(d\lambda)$. However, this is the same as dragging along of $\Phi_\Lambda$ over $v^m(x^k)d\lambda$ by $T^{-1}(d\lambda)$. It thus follows that the condition $\underset{v}{\Delta}\Phi_\Lambda = 0$ is equivalent to the evaluation of the variation in $\{\Phi_\Lambda\}$ when the points are moved around by a point transformation. This may seem strange at first, but it is really what should be expected in view of the relation between point transformations and coordinate transformations that obtain by the requirement that the point transformation drag along the coordinate transformation. Also, it is obvious that, for point transformations, there must be function changes in addition to those which arise from coordinate transformations. This is required since the field $\{\Phi_\Lambda\}$ is being evaluated at new points under a point transformation, while a coordinate transformation leaves the point of evaluation fixed and refers the field to a new base defined by the coordinate transformation. We shall thus be able to consider both point transformations and coordinate transformations within the same notation. From (4.11), (4.12) and (4.17), we also have

$$\begin{aligned}\frac{\partial\Phi_{\Lambda'}(x; \lambda, \eta)}{\partial x^{k'}} &\equiv \widetilde{\partial_k \Phi_{\Lambda'}}(x) \\ &= \partial_k\Phi_\Lambda + \eta\partial_k(\delta\Phi_\Lambda) - \lambda[\partial_k v^m \partial_m\Phi_\Lambda - \partial_k(\underset{v}{D}\Phi_\Lambda)] + o(\lambda, \eta) .\end{aligned} \tag{4.19}$$

Thus, the $N(n+1)$ functions $\{\widetilde{{}_{\alpha}\Phi_{\Lambda'}(x)}\}$ are well defined.

We have assumed that the q functions $g_a$ and the function $L$ are arbitrary but fixed functions of their indicated arguments. We may accordingly form the functionals

$$\tilde{k}_a(x^{m'}, \{\Phi_\Lambda\}; d) = \int_d g_a[x^{m'}, z^{m'}, \widetilde{{}_{\alpha}\Phi_{\Lambda'}}(z')]\, dV(z') \tag{4.20}$$

and

$$\tilde{J}[\{\Phi_\Lambda\}](L; d) = \int_d L[x^{m'}, \widetilde{{}_{\alpha}\Phi_{\Lambda'}}(x^m), \tilde{k}_a(x^{m'}, \{\Phi_\Lambda\}; d)]\, dV(x')\ . \tag{4.21}$$

As an example, suppose that

$$g_1[x^m, z^m, {}_{\alpha}\Phi_\Lambda(z)] = H^{\Lambda}(x^m, z^m)\,\Phi_\Lambda(z^m)$$

and that $\Phi_{\Lambda'}(x) = F_\Lambda{}^{\Gamma}(x^m, x^{m'})\,\Phi_\Gamma(x)$; then

$$\begin{aligned} g_1[x^{m'}, z^{m'}, \widetilde{{}_{a}\Phi_{\Lambda'}}(z)] &= H^{\Lambda}(x^{m'}, z^{m'})\,\widetilde{\Phi_{\Lambda'}}(z^m) \\ &= H^{\Lambda}(x^{m'}, z^{m'})\,F_\Lambda{}^{\Gamma}(z^m, z^{m'})\,\widetilde{\Phi_\Gamma}(z^m)\ . \end{aligned}$$

Consequently

$$\tilde{k}_a(x^{m'}, \{\Phi_\Lambda\}; d) = \int_d H^{\Lambda}(x^{m'}, z^{m'})\,F_\Lambda{}^{\Gamma}(x^m, z^{m'})\,\widetilde{\Phi_\Gamma}(z^m)\, dV(z')$$

where $z^m$ is to be expressed in terms of $z^{m'}$ by the transformation equations $z^m = f^m(z^{1'}, \ldots, z^{n'})$.

From this example and similar ones constructed using $L$ rather than $g_a$, it is evident that $\tilde{k}_a$ differs from $k_a$ and that $\tilde{J}$ differs from $J$ in general. The difference

$$\Delta J[\{\Phi_\Lambda\}](L; d) = \tilde{J}[\{\Phi_\Lambda\}](L; d) - J[\{\Phi_\Lambda\}](L; d) \tag{4.22}$$

is referred to as the *extended finite variation of* $J$. Written out in full, (4.22) becomes

$$\begin{aligned} \Delta J[\{\Phi_\Lambda\}](L, d) = &\int_d L[x^{m'}, \widetilde{{}_{\alpha}\Phi_{\Lambda'}}(x^m), \tilde{k}_a(x^{m'}, \{\Phi_\Lambda\}; d)]\, dV(x') \\ &- \int_d L[x^m, {}_{\alpha}\Phi_\Lambda(x^m), k_a(x^m, \{\Phi_\Lambda\}; d)]\, dV(x)\ . \end{aligned} \tag{4.23}$$

If we transform the first integral on the right-hand side of (4.23) into an integration with respect to the coordinates $x^m$ and if we set

$$\frac{\partial(x')}{\partial(x)} \equiv \left|\det\left(\frac{\partial x^{m'}}{\partial x^k}\right)\right| \equiv \left(\frac{x'}{x}\right), \tag{4.24}$$

we then have

$$\Delta J[\{\Phi_\Lambda\}](L;d) = \int_d \Big\{ L[x^{m'}, \widetilde{{}_a\Phi_{\Lambda'}}(x^m), \tilde{k}_a(x^{m'}, \{\Phi_\Lambda\}; d)] \frac{\partial(x')}{\partial(x)} \tag{4.25}$$

$$- L[x^m, {}_a\Phi_\Lambda(x^m), k_a(x^m, \{\Phi_\Lambda\}; d]\Big\} dV(x) .$$

Further considerations will be simplified if we first establish several preliminary results.

**Lemma 4.1.** *If the coordinates on* D* *are transformed in accordance with* (4.11) *and* (4.13), *we have*

$$\frac{\partial(x')}{\partial(x)} = 1 + \lambda \partial_k v^k(x^m) + o(\lambda) . \tag{4.26}$$

**Proof.** When (4.11) and (4.12) are combined, we obtain

$$\frac{\partial x^{m'}}{\partial x^k} = \delta_k{}^m + \lambda \partial_k v^m(x^t) + o(\lambda) ,$$

$$\det[\delta_k{}^m + \lambda\partial_k v^m + o(\lambda)] = \lambda^n \det\left[\frac{1}{\lambda}\delta_k{}^m + \partial_k v^m + \frac{o(\lambda)}{\lambda}\right]$$

$$= \lambda^n \left\{\left(\frac{1}{\lambda}\right)^n + \left(\frac{1}{\lambda}\right)^{n-1} \operatorname{trace}(\partial_k v^m) + o(\lambda^{1-n})\right\}$$

$$= 1 + \lambda \operatorname{trace}(\partial_k v^m) + o(\lambda) .$$

Substituting these results into (4.24), we establish the relation (4.26). We say that a transformation is volume preserving if $\partial_k v^k = 0$.

**Lemma 4.2.** *Under the coordinate transformations* (4.11) *to* (4.13) *and the function variations* (4.6), *we have*

$$\tilde{k}_a(x^{m'}, \{\Phi_\Lambda\}; d) = k_a(x^m, \{\Phi_\Lambda\}; d) + o(\eta, \lambda) + \lambda \int_d \Big\{ \frac{\partial v^m(z)}{\partial z^m} g_a +$$

$$+ v^m(x)\frac{\partial^* g_a}{\partial x^m} + v^m(z)\frac{\partial^* g_a}{\partial z^m} + \underset{v}{D}\Phi_\Lambda(z)\frac{\partial g_a}{\partial \Phi_\Lambda}$$

$$+ \left[\frac{\partial}{\partial z^k}(\underset{v}{D}\Phi_\Lambda(z)) - \frac{\partial v^m(z)}{\partial z^k}\frac{\partial \Phi_\Lambda(z)}{\partial z^m}\right]\frac{\partial g_a}{\partial[\partial_k \Phi_\Lambda(z)]}\Bigg\} dV(z)$$

$$+ \eta \int_d \left\{\delta\Phi_\Lambda(z)\frac{\partial g_a}{\partial \Phi_\Lambda} + \frac{\partial \delta\Phi_\Lambda(z)}{\partial z^k}\frac{\partial g_a}{\partial[\partial_k \Phi_\Lambda(z)]}\right\} dV(z) . \tag{4.27}$$

**Proof.** When (4.11), (4.12), (4.17) and (4.19) are substituted into (4.20) and the integration with respect to $(x')$ is converted into an integration with respect to $(x)$, we have

$$\tilde{k}_a(x^{m'}, \{\Phi_\Lambda\}; d) = \int_d \left[1 + \lambda\frac{\partial v^m(z)}{\partial z^m} + o(\lambda)\right] g_a\Big\{x^m + \lambda v^m(x) + o(\lambda),$$

$$z^m + \lambda v^m(z) + o(\lambda), \Phi_\Lambda(z) + \lambda \underset{v}{D}\Phi_\Lambda(z) + \eta\delta\Phi_\Lambda(z)$$

$$+ o(\eta, \lambda), \frac{\partial \Phi_\Lambda(z)}{\partial z^k} + \eta\frac{\partial}{\partial z^k}[\delta\Phi_\Lambda(z)] - \lambda\left[\frac{\partial v^m(z)}{\partial z^k}\frac{\partial \Phi_\Lambda(z)}{\partial z^m}\right.$$

$$\left.\left. - \frac{\partial}{\partial z^k}(\underset{v}{D}\Phi_\Lambda(z))\right]\right\} dV(z) ,$$

$$= k_a(x^m, \{\Phi_\Lambda\}; d) + \int_d \Big\{\lambda\frac{\partial v^m(z)}{\partial z^m} g_a + \lambda v^m(x)\frac{\partial^* g_a}{\partial x^m}$$

$$+ \lambda v^m(z)\frac{\partial^* g_a}{\partial z^m} + [\lambda \underset{v}{D}\Phi_\Lambda(z) + \eta\delta\Phi_\Lambda(z)]\frac{\partial g_a}{\partial \Phi_\Lambda}$$

$$+ \left[\eta\frac{\partial \delta\Phi_\Lambda(z)}{\partial z^k} - \lambda\frac{\partial v^m(z)}{\partial z^k}\frac{\partial \Phi_\Lambda(z)}{\partial z^m} + \lambda\frac{\partial}{\partial z^k}(\underset{v}{D}\Phi_\Lambda(z))\right] \cdot$$

$$\cdot \frac{\partial g_a}{\partial(\partial_k \Phi_\Lambda)}\Big\} dV(z) + o(\lambda, \eta) .$$

Collecting the various terms together, we obtain (4.27).

**Lemma 4.3.** *The system of relations* (4.27) *is equivalent to the system*

$$\tilde{k}_a(x^{m'}, \{\Phi_\Lambda\}; d) = k_a(x^{m'}, \{\Phi_\Lambda\}; d) + \Delta k_a(x^m, \{\Phi_\Lambda\}; d) + o(\lambda, \eta) \tag{4.28}$$

*where*

$$\Delta k_a(x^m, \{\Phi_\Lambda\}; d) = \int_d \frac{\partial}{\partial z^m}\left\{\lambda\left[\delta_k{}^m g_a - \frac{\partial g_a}{\partial[\partial_m \Phi_\Lambda(z)]}\frac{\partial \Phi_\Lambda(z)}{\partial z^k}\right]v^k(z) + \frac{\partial g_a}{\partial[\partial_m \Phi_\Lambda(z)]}[\lambda \underset{v}{D}\Phi_\Lambda(z) + \eta\delta\Phi_\Lambda(z)]\right\}dV(z)$$
$$+ \int_d \{e|g_a\}_{\Phi_\Lambda}[\eta\delta\Phi_\Lambda(z) - \lambda \underset{v}{\pounds}\Phi_\Lambda(z)]\,dV(z)$$
$$+ \int_d v^m(x)\frac{\partial^* g_a}{\partial x^m}\,dV(z)\,. \tag{4.29}$$

**Proof.** When (4.15) is substituted into the integrand appearing in (4.27), the integrand becomes

$$\lambda\left\{\frac{\partial v^m(z)}{\partial z^m}g_a + v^m(x)\frac{\partial^* g_a}{\partial x^m} + v^m(z)\frac{\partial^* g_a}{\partial z^m} + \left[v^m(z)\frac{\partial \Phi_\Lambda}{\partial z^m} - \underset{v}{\pounds}\Phi_\Lambda(z)\right]\frac{\partial g_a}{\partial \Phi_\Lambda}\right.$$
$$\left. + \frac{\partial}{\partial z^k}\left[v^m(z)\frac{\partial \Phi_\Lambda(z)}{\partial z^m} - \underset{v}{\pounds}\Phi_\Lambda(z)\right]\frac{\partial g_a}{\partial[\partial_k \Phi_\Lambda(z)]} - \frac{\partial v^m(z)}{\partial z^k}\frac{\partial \Phi_\Lambda(z)}{\partial z^m}\frac{\partial g_a}{\partial[\partial_k \Phi_\Lambda(z)]}\right\}$$
$$+ \eta\left\{\delta\Phi_\Lambda(z)\frac{\partial g_a}{\partial \Phi_\Lambda(z)} + \frac{\partial \delta\Phi_\Lambda(z)}{\partial z^k}\frac{\partial g_a}{\partial[\partial_k \Phi_\Lambda(z)]}\right\}$$
$$= \frac{\partial}{\partial z^m}\left\{\lambda g_a v^m + \frac{\partial g_a}{\partial[\partial_m \Phi_\Lambda(z)]}[\eta\delta\Phi_\Lambda(z) - \lambda\underset{v}{\pounds}\Phi_\Lambda(z)]\right\} + v^m(x)\frac{\partial^* g_a}{\partial x^m}$$
$$+ \{e|g_a\}_{\Phi_\Lambda}[\eta\delta\Phi_\Lambda(z) - \lambda\underset{v}{\pounds}\Phi_\Lambda(z)]\,.$$

Hence, when (4.15) is used again, so that

$$-\frac{\partial g_a}{\partial[\partial_k \Phi_\Lambda(z)]}\underset{v}{\pounds}\Phi_\Lambda(z) = \frac{\partial g_a}{\partial[\partial_k \Phi_\Lambda(z)]}\left[\underset{v}{D}\Phi_\Lambda - v^m(z)\frac{\partial \Phi_\Lambda(z)}{\partial z^m}\right],$$

we obtain (4.29).

**Lemma 4.4.** *Under the coordinate transformations* (4.11) *to* (4.13) *and the function variations* (4.6), *we have*

$$L[x^{m'}, {}_{\alpha}\Phi_{\Lambda'}(x), \tilde{k}_a(x^{m'}, \{\Phi_\Lambda\}; d)]\frac{\partial(x')}{\partial(x)} =$$

$$= L[x^m, {}_\alpha\Phi_\Lambda(x), k_a(x^m, \{\Phi_\Lambda\}; d)] + \frac{\partial}{\partial x^m}\left\{\lambda\left[\delta_k{}^m L - \frac{\partial L}{\partial(\partial_m\Phi_\Lambda)}\frac{\partial\Phi_\Lambda}{\partial x^k}\right]v^k\right.$$

$$+ \left.\frac{\partial L}{\partial(\partial_m\Phi_\Lambda)}(\lambda \underset{v}{D}\Phi_\Lambda + \eta\delta\Phi_\Lambda)\right\} + \{e|L(k_a)\}_{\Phi_\Lambda}(x^m)(\eta\delta\Phi_\Lambda - \lambda \underset{v}{\pounds}\Phi_\Lambda)$$

$$+ \frac{\partial L}{\partial k_a}\left\{\int_d \frac{\partial}{\partial z^m}\left\{\lambda\left[\delta_k{}^m g_a - \frac{\partial g_a}{\partial[\partial_m\Phi_\Lambda(z)]}\frac{\partial\Phi_\Lambda(z)}{\partial z^k}\right]v^k(z)\right.\right.$$

$$+ \left.\frac{\partial g_a}{\partial[\partial_m\Phi_\Lambda(z)]}[\lambda \underset{v}{D}\Phi_\Lambda(z) + \eta\delta\Phi_\Lambda(z)]\right\}dV(z) + \int_d \{e|g_a\}_{\Phi_\Lambda}(z^m)\,\cdot$$

$$\left.\cdot\,[\eta\delta\Phi_\Lambda(z) - \lambda \underset{v}{\pounds}\Phi_\Lambda(z)]\,dV(z)\right\} + o(\eta, \lambda)\,. \tag{4.30}$$

**Proof.** This result is obtained by exactly the same method as used in the Lemmas 4.2 and 4.3 and is left to the student as an exercise.

**Lemma 4.5.** *Under the coordinate transformations* (4.11) *to* (4.13) *and the function variations* (4.6), *we have*

$$\Delta J[\{\Phi_\Lambda\}](L; d) = \int_d \{\mathcal{E}|L\}_{\Phi_\Lambda}[\eta\delta\Phi_\Lambda - \lambda \underset{v}{\pounds}\Phi_\Lambda(x)]\,dV(x) + \int_d \frac{\partial}{\partial x^m}\left\{-\lambda W_k{}^m v^k\right.$$

$$+ \left[\frac{\partial L}{\partial(\partial_m\Phi_\Lambda)} + \int_d \frac{\partial L}{\partial k_a}(z)\frac{\partial g_a^*}{\partial[\partial_m\Phi_\Lambda(x)]}\,dV(z)\right][\lambda \underset{v}{D}\Phi_\Lambda(x)$$

$$\left. + \eta\delta\Phi_\Lambda(x)]\right\}dV(x) + o(\eta, \lambda)\,, \tag{4.31}$$

*in which* $(W_k^m)$ *is the momentum-energy complex of* $L$.

**Proof.** We first substitute (4.30) into (4.25). If the order of integration is reversed in the resulting double integrals ($(x^k, z^k)$ are replaced by $(z^k, x^k)$, which is equivalent to having terms of the form $(\partial L/\partial k_a)(z)\,\partial g_a^*$ instead of $(\partial L/\partial k_a)\,\partial g_a$) and the definitions of $\{\mathcal{E}|L\}$ and the momentum-energy complex are used, a little work yields (4.31)

**Theorem 4.1.** *Let* $L(x^m, {}_a\Phi_\Lambda(x^m), k_a(x^m, \{\Phi_\Lambda\}, d))$ *be a given lagrangian function whose arguments undergo the transformation* (4.11) *to* (4.13), (4.17), (4.19), (4.28) *and* (4.20); *then, for the extended variations of* $J[\{\Phi_\Lambda\}](L; d)$ *which result, we have*

$$\left.\frac{dJ}{d\lambda}\right|_{\lambda=\eta=0}(d) = \int_d \frac{\partial}{\partial x^m}\left\{-W_k{}^m v^k + \left[\frac{\partial L}{\partial(\partial_m \Phi_\Lambda)} + \int_d \frac{\partial L}{\partial k_a}(z) \cdot \frac{\partial g_a^*}{\partial[\partial_m \Phi_\Lambda(x)]}\right] \underset{v}{\mathfrak{D}} \Phi_\Lambda\right\} dV(x) - \int_d \{\mathcal{E}|L\}_{\Phi_\Lambda} \underset{v}{\pounds} \Phi_\Lambda dV(x), \tag{4.32}$$

$$\left.\frac{dJ}{d\eta}\right|_{\eta=\lambda=0}(d) = \int_d \{\mathcal{E}|L\}_{\Phi_\Lambda} \delta\Phi_\Lambda dV(x) + \int_d \frac{\partial}{\partial x^m}\left\{\left[\frac{\partial L}{\partial(\partial_m \Phi_\Lambda)} + \int_d \frac{\partial L}{\partial k_a}(z) \frac{\partial g_a^*}{\partial[\partial_m \Phi_\Lambda(x)]} dV(z)\right] \delta\Phi_\Lambda(x)\right\} dV(x). \tag{4.33}$$

*Also, under the confluence* $\lambda = \pi = \eta$,

$$\left.\frac{dJ}{d\pi}\right|_{\pi=0}(d) = \int_d \{\mathcal{E}|L\}_{\Phi_\Lambda}(\delta\Phi_\Lambda - \underset{v}{\pounds}\Phi_\Lambda) dV(x) + \int_d \frac{\partial}{\partial x^m}\left\{-W_k{}^m v^k + \left[\frac{\partial L}{\partial(\partial_m \Phi_\Lambda)} + \int_d \frac{\partial L}{\partial k_a}(z) \frac{\partial g_a^*}{\partial(\partial_m \Phi_\Lambda)} dV(z)\right] \underset{v}{\Delta} \Phi_\Lambda\right\} dV(x). \tag{4.34}$$

**Proof.** The difficult work has been done, and all that remains is to use Lemma 4.5 and note that

$$\left.\frac{dJ}{d\lambda}\right|_{\lambda=\eta=0}(d) = \lim_{\lambda\to 0}\left[\frac{\Delta J[\{\Phi_\Lambda\}](L;d)|_{\eta=0}}{\lambda}\right], \quad \text{etc.}$$

The student should note that $dJ/d\eta|_{\eta=\lambda=0}$ is another way of writing $\delta J$. We consequently refer to $dJ/d\lambda|_{\lambda=\eta=0}$ as the *coordinate variation* $\delta J/\delta x$ since this variation in $J$ arises from a coordinate transformation. We also refer to $dJ/d\eta|_{\lambda=\eta=0}$ as the *function variation* $\delta J/\delta\{\Phi_\Lambda\}$ since it arises from a function variation. Similarly, we refer to $dJ/d\pi|_{\pi=0}$ as the *total variation* $\delta J/\Delta\{\Phi_\Lambda\}$. Accordingly, we say that $J$ *is stationary with respect to coordinate variations* if $\delta J/\delta x = 0$, that $J$ *is stationary with respect to function variations* if $dJ/d\eta = 0$ and that $J$ *is stationary with respect to total variations* if $dJ/d\pi = 0$.

Up to this point, we have considered the general situation in which the lagrangian function is itself a function of functionals, and hence the Euler equations are generally integrodifferential equations.

In order to simplify the analyses of the following sections and to reduce the problems to those usually considered in the literature, we make the following agreement: *there are no functional arguments of the lagrangian functions, in other words,*

$$\frac{\partial L}{\partial k_a} = 0 \tag{4.35}$$

*for all lagrangian functions under consideration.* Under this agreement, we have the following simplications and results:

$$\{\mathcal{E}|L\}_{\Phi_\Lambda} \equiv \{e|L\}_{\Phi_\Lambda}, \tag{4.36}$$

$$W_k{}^m(L) = \frac{\partial L}{\partial(\partial_m \Phi_\Lambda)} \partial_k \Phi_\Lambda - \delta_k{}^m L, \tag{4.37}$$

the strong identity

$$\partial_m W_k{}^m(L) = -\{\mathcal{E}|L\}_{\Phi_\Lambda} \partial_k \Phi_\Lambda - \partial_k^* L, \tag{4.38}$$

the weak identity

$$\partial_m W_k{}^m(L) \underset{S(L)}{=} -\partial_k^* L, \tag{4.39}$$

$$\begin{aligned} \left.\frac{dJ}{d\lambda}\right|_{\lambda=\eta=0}(d) &= \int_d \frac{\partial}{\partial x^m}\left\{-W_k{}^m v^k + \frac{\partial L}{\partial(\partial_m \Phi_\Lambda)} \underset{v}{D}\Phi_\Lambda\right\} dV(x) \\ &\quad - \int_d \{\mathcal{E}|L\}_{\Phi_\Lambda} \underset{v}{\pounds}\Phi_\Lambda \, dV(x), \end{aligned} \tag{4.40}$$

$$\left.\frac{dJ}{d\eta}\right|_{\eta=\lambda=0}(d) = \int_d \{\mathcal{E}|L\}_{\Phi_\Lambda} \delta\Phi_\Lambda \, dV(x) + \int_d \frac{\partial}{\partial x^m}\left\{\frac{\partial L}{\partial(\partial_m \Phi_\Lambda)} \delta\Phi_\Lambda\right\} dV(x), \tag{4.41}$$

$$\begin{aligned} \left.\frac{dJ}{d\pi}\right|_{\pi=0}(d) &= \int_d \{\mathcal{E}|L\}_{\Phi_\Lambda} (\delta\Phi_\Lambda - \underset{v}{\pounds}\Phi_\Lambda) \, dV(x) \\ &\quad + \int_d \frac{\partial}{\partial x^m}\left\{-W_k{}^m v^k + \frac{\partial L}{\partial(\partial_m \Phi_\Lambda)} \underset{v}{\Delta}\Phi_\Lambda\right\} dV(x), \end{aligned} \tag{4.42}$$

where

$$\underset{v}{D}\Phi_\Lambda(x^m) = v^k \partial_k \Phi_\Lambda - \underset{v}{\pounds}\Phi_\Lambda \tag{4.43}$$

and

$$\underset{v}{\Delta}\Phi_\Lambda(x) = \underset{v}{D}\Phi_\Lambda(x^m) + \delta\Phi_\Lambda(x^m) \tag{4.44}$$

is the total infinitesimal variation of $\Phi_\Lambda$ that results under the confluence $\eta = \pi = \lambda$ (which results under a composition of functional variation and coordinate changes $x^{m'} = x^m + \pi v^m(x^k) + o(\pi)$, $\tilde{\Phi}_{\Lambda'}(x^m) = \Phi_\Lambda(x^m) + \pi[\underset{v}{\Delta}\Phi_\Lambda(x^m)] + o(\pi)$).

The forms presented above have had their terms collected together in a very special way so as to involve the Euler–Lagrange operator and $W_k{}^m$. If this is not done, we have the equivalent forms

$$\left.\frac{dJ}{d\lambda}\right|_{\lambda=\eta=0} = \int_d \left\{ L\partial_m v^m + v^m \partial^*_m L + \underset{v}{D}\Phi_\Lambda \frac{\partial L}{\partial \Phi_\Lambda} + (\partial_m \underset{v}{D}\Phi_\Lambda - \partial_m v^k \partial_k \Phi_\Lambda)\frac{\partial L}{\partial(\partial_m \Phi_\Lambda)}\right\} dV(x)\,, \tag{4.45}$$

$$\left.\frac{dJ}{d\eta}\right|_{\lambda=\eta=0} = \int_d \left\{ \delta\Phi_\Lambda \frac{\partial L}{\partial \Phi_\Lambda} + \partial_m \delta\Phi_\Lambda \frac{\partial L}{\partial(\partial_m \Phi_\Lambda)}\right\} dV(x)\,, \tag{4.46}$$

$$\left.\frac{dJ}{d\pi}\right|_{\pi=0} = \int_d \left\{ L\partial_m v^m + v^m \partial^*_m L + \underset{v}{\Delta}\Phi_\Lambda \frac{\partial L}{\partial \Phi_\Lambda} + (\partial_m \underset{v}{\Delta}\Phi_\Lambda - \partial_m v^k \partial_k \Phi_\Lambda)\frac{\partial L}{\partial(\partial_m \Phi_\Lambda)}\right\} dV(x)\,. \tag{4.47}$$

## 4.3. WEAK COORDINATE INVARIANCE

For any $d \subset D^*$ and for any given lagrangian function $L$, we form the functional $J[\{\Phi_\Lambda\}](L;d)$ on the space $\mathfrak{D}_1(D^*;N)$ of all $N$-tuples of $C^1$ functions $\{\Phi_\Lambda(x^m)\}$ that are defined over the $n$-dimensional region $D^*$ of $X_n{}^u$, when this region $D^*$ is referred to a given coordinate system $(x)$. Let us introduce a one-parameter family of coordinate transformations by the relations (4.11) to (4.13) where the $n$ function $v^k(x^m)$ are explicit, known functions of their arguments for all points in $D^*$. This family of coordinate transformations serves to define a one-parameter family of coordinate systems $(x')$ on $D^*$, and we may accordingly refer $D^*$ to any coordinate system of this family. In actuality, the equations (4.11) to (4.13) define a one-parameter continuous group, $G_1$, of coordinate transformations and a one-parameter group of coordinate systems over $D^*$ such that when $\lambda = 0$, the $(x')$ -coordinate system and the $(x)$ coordinate system coincide. This group $G_1$ is a subgroup of the pseudogroup $G_n{}^u$, and is obviously a proper subgroup since $G_n{}^u$ contains the group of all analytic coordinate transformations which is not a finite parameter group. The functionals $\tilde{J}[\{\Phi_{\Lambda'}\}](L;d)$ are then formed on the space $\mathfrak{D}_1(D^*;N)$ of all $N$-tuples of $C^1$ functions $\{\Phi_{\Lambda'}(x^m)\}$ defined over the $n$-dimensional systems over $D^*$. Since we have assumed that $\{\Phi_\Lambda(x^m)\}$ forms a geometric object field, then the $\{\Phi_\Lambda(x^m)\}$ and the $\{\Phi_{\Lambda'}(x^m)\}$ are in a one-to-one correspondence (see Chap. 3).

Now, we have used the same function $L$ in forming both $J$ and $\tilde{J}$. Hence, if the values of $J$ and $\tilde{J}$ are different for any given $\{\Phi_\Lambda(x^m)\} \in \mathfrak{D}_1(D^*;N)$, this difference arises because we have used different coordinates on $D^*$. It is assumed here that if we use the element $\{\Phi_\Lambda(x^m)\}_0$ in evaluating $J$, then, for evaluation $\tilde{J}$, we use the image $\{\Phi_{\Lambda'}(x^m)\}_0$ of $\{\Phi_\Lambda(x^m)\}_0$ (which results when the geometric object field $\{\Phi_\Lambda(x^m)\}_0$ is subjected to the transformation $\Phi_{\Lambda'} = F_\Lambda(\Phi_{\Sigma 0}, x^m, x^{m'})$ induced by the transformation of coordinates on $D^*$.) It thus follows by excluding function variations from the considerations of Sec. 4.2 (that is, $\eta \equiv 0$) that the difference

$$\Delta J[\{\Phi_\Lambda\}](L;d) = \tilde{J}[\{\Phi_\Lambda\}](L;d) - J[\{\Phi_\Lambda\}](L;d) \tag{4.48}$$

is just the change in the value of $J[\{\Phi_\Lambda\}](L;d)$ that arises from changing the coordinate system over $D^*$ by a member of the one-parameter group $G_1$ of coordinate transformations (4.11) to (4.13) with infinitesimal generating vector $v^k(x^m)$.

The reason for considering subdomains $d$ of $D$ is now apparent. Suppose that we had considered only the domain $D$. In that case, if the quantity given by (4.48) with $d = D$ vanishes, the value of the functional $J[\{\Phi_\Lambda\}](L;D)$ would be unchanged for coordinate transformations generated by the one-parameter group $G_1$, and hence the idea of invariance under $G_1$ would be intimately connected with the domain $D$. Such invariance under coordinate transformations would be of a composite nature and add further complication since there would, in general, be a lack of local coordinate invariance—the non-invariance of one part of $D$ could exactly balance the lack of invariance of another part of $D$ so that the total integrated effect would be zero. This could possibly be allowed if there were some intrinsic method of singling out the domain $D$; however, in most applications the basic domain $D$ is chosen more for convenience than for any other reason and is not intrinsic to the problem. On the other hand, if the quantity given by (4.48) vanishes for all $d \subset D$, then the value of the functional $J[\{\Phi_\Lambda\}](L, D)$ is certainly unchanged in value under coordinate transformations of the one-parameter group $G_1$. In addition, this property is also shared for every subdomain and hence is intrinsically associated with the lagrangian function that defines the functional, not just with the integral of this function. Since it is the lagrangian function which basically defines the properties of the functional under investigation and which represents the basic physics in those cases containing an underlying physical model, characterization of notions of invariance should rest on the lagrangian function itself rather than on the integral of this function over some fixed domain $D$. For these reasons we make the following definition: *A functional $J[\{\Phi_\Lambda\}](L, D)$ is said to be weakly coordinate invariant under a one-parameter group $G_1$ with infinitesimal generator $v^k(x^m)$ if and only if $\Delta J[\{\Phi_\Lambda\}](L, d)$ vanishes for all $d$ in $D$ with $\delta\Phi_\Lambda \equiv 0$.* The reason for referring to this invariance as week invariance is that we are speaking of the invariance of the value of $J[\{\Phi_\Lambda\}](L; d)$ so that we are dealing with a fixed element $\{\Phi_\Lambda(x^m)\}$ of $\mathfrak{D}_1(D^*; N)$. Later, we shall consider strong coordinate invariance in which the functional itself is invariant, i.e., the functional viewed as a mapping of $\mathfrak{D}_1(D^*; N)$ into the reals is an invariant mapping under coordinate transformations for strong coordinate invariance.

The results obtained in Sec. 4.2 provide us with the elements required to establish conditions for weak coordinate invariance. From the definition, we must have $\Delta J[\{\Phi_\Lambda\}](L; d) = 0$ for all $d \subset D$ and for $\delta\Phi_\Lambda \equiv 0$; and hence the change in the value of $J[\{\Phi_\Lambda\}](L; d)$ must vanish when the coordinate transformation is restricted to a sufficiently small neighborhood of the identity transformation for all $d \subset D$ and for $\delta\Phi_\Lambda \equiv 0$. We must accordingly have $dJ/d\lambda|_{\lambda=\eta=0} = 0$ for all $d \subset D$. When equation (4.40) is used, we then establish the following result.

**Theorem 4.2.** *Let $L$ be a given lagrangian function and let $\Phi_\Lambda(x^m)$ be a given $N$-tuple of functions that constitutes and element of $\mathfrak{D}_1(D^*; N)$. If $J[\{\Phi_\Lambda\}](L; D)$ is weakly coordinate invariant under the one-parameter group (germ)*

$$G_1 : x^{m'} = G^m(x^k, \lambda) = x^m + \lambda v^m(x^k) + o(\lambda) \tag{4.49}$$

*of coordinate transformations, then the quantities $\{\Phi_\Lambda(x^m)\}$ and $v^m(x^k)$ must satisfy the relation*

$$\frac{\partial}{\partial x^m}\left\{\frac{\partial L}{\partial(\partial_m \Phi_\Lambda)} \underset{v}{D} \Phi_\Lambda - W_k{}^m v^k\right\} = \{\mathcal{E}|L\}_{\Phi_\Lambda} \underset{v}{\pounds} \Phi_\Lambda . \tag{4.50}$$

*If* $\{\Phi_\Lambda(x^m)\}$ *and* $v^m(x^k)$ *satisfy the relation* (4.50), *then* $J[\{\Phi_\Lambda\}](L, D)$ *is weakly coordinate invariant under the one-parameter group* $G_1$ *of coordinate transformations restricted to a sufficiently small neighborhood of the identity transformation. Further, if* $J[\{\Phi_\Lambda\}](L, D)$ *is weakly coordinate invariant under an* $r$*-parameter continuous group* $G_r$ *of coordinate transformations with infinitesimal generators* $\underset{a}{v}{}^m(x^k)$, $a = 1, \ldots, r$, *then* $\{\Phi_\Lambda(x^m)\}$ *and* $\underset{a}{v}{}^m(x^k)$ *must satisfy the* $r$ *relations*

$$\frac{\partial}{\partial x^m}\left\{\frac{\partial L}{\partial(\partial_m \Phi_\Lambda)} \underset{v_a}{D} \Phi_\Lambda - W_k{}^m \underset{a}{v}{}^k\right\} = \{\mathcal{E}|L\}_{\Phi_\Lambda} \underset{v_a}{\pounds} \Phi_\Lambda . \tag{4.51}$$

*If the relations* (4.51) *hold, then* $J[\{\Phi_\Lambda\}](L, D)$ *is weakly coordinate invariant under the* $r$*-parameter group* $G_r$ *of coordinate transformations restricted to a sufficiently small neighborhood of the identity.*

The student should clearly note that Theorem 4.2 gives partial differential equations for the determination of the infinitesimal generators of the group for which $J[\{\Phi_\Lambda\}](L, D)$ is invariant under coordinate substitutions. This follows from the fact that $\{\Phi_\Lambda(x^m)\}$ is assumed to be a given element of $\mathfrak{D}_1(D^*, N)$ and hence constitutes an $N$-tuple of known functions of $x^m$ for all $x^m$ in $D^*$.

A number of important corollaries follow immediately from Theorem 4.2 and from the results of Sec. 4.2.

*If* $\{\Phi_\Lambda(x^m)\}$ *is an element of* $S(L)$, *then weak coordinate invariance of* $J[\{\Phi_\Lambda\}](L, D)$ *under a* $G_r$ *of coordinate transformations implies the* $r$ *conservation laws*

$$\frac{\partial}{\partial x^m}\left\{\frac{\partial L}{\partial(\partial_m \Phi_\Lambda)} \underset{v_a}{D} \Phi_\Lambda - W_k{}^m \underset{a}{v}{}^k\right\} = 0 . \tag{4.52}$$

*Conversely, if the* $r$ *conservation laws* (4.52) *hold and* $\{\Phi_\Lambda\} \in S(L)$, *then the value of* $J[\{\Phi_\Lambda\}](L, D)$ *is unchanged under coordinate transformations of the group* $G_r$ *when the group is restricted to a sufficiently small neighborhood of the identity transformation.*

*If* $G_r$ *is an invariance group of* $\{\Phi_\Lambda(x^m)\}$ *(that is,* $\underset{v}{\pounds}\Phi_\Lambda = 0$*), then weak coordinate invariance of* $J[\{\Phi_\Lambda\}](L, D)$ *under* $G_r$ *gives the* $r$ *conservation laws* (4.52). *Conversely, if the* $r$ *conservation laws* (4.52) *hold and* $G_r$ *is an invariance group of* $\{\Phi_\Lambda(x^m)\}$, *then the value* *of* $J[\{\Phi_\Lambda\}](L,D)$ *is unchanged under coordinate transformations of the group in a sufficiently small neighborhood of the identity transformation of* $G_r$*. In this instance, since*

$$\underset{v_a}{D}\Phi_\Lambda = v_a{}^k \partial_k \Phi_\Lambda - \underset{v_a}{\pounds}\Phi_\Lambda ,$$

*we have* $\underset{v_a}{D}\Phi_\Lambda = v_a^k \partial_k \Phi_\Lambda$, *and* (4.52) *reduces to*

$$\frac{\partial}{\partial x^m}\left[\left\{\frac{\partial L}{\partial(\partial_m \Phi_\Lambda)}\partial_k \Phi_\Lambda - W_k{}^m\right\} v_a{}^k\right] = \frac{\partial}{\partial x^m}[L v_a{}^m] = 0 . \tag{4.53}$$

Suppose that we have an $N$-tuple of functions $\{\Phi_\Lambda(x^m)\}$ which constitute an element of $\mathfrak{D}_1(D^*;N)$ and an $r$-parameter group $G_r$ of coordinate transformations with infinitesimal generators $v_a{}^m(x^k)$ such that the $r$ conservation laws (4.52) are satisfied. Can we conclude that $\{\Phi_\Lambda(x^m)\}$ is an element of $S(L)$ if the value of $J[\{\Phi_\Lambda\}](L,D)$ is invariant in value under coordinate transformations of the group $G_r$ restricted to a sufficiently small neighborhood of the identity transformation? The answer is obviously no, since an inspection of (4.50) shows that all of the hypotheses are fulfilled if $G_r$ is an invariance group of $\{\Phi_\Lambda(x^m)\}$. Even if this possibility is eliminated, the above conclusion still does not follow. *If the infinitesimal generators of an* $r$*-parameter group* $G_r$ *and an element* $\{\Phi_\Lambda(x^m)\}$ *of* $\mathfrak{D}_1(D^*;N)$ *can be found such that the conservation laws* (4.50) *are satisfied and if* $J[\{\Phi_\Lambda\}](L;D)$ *is weakly coordinate invariant under* $G_r$, *then* $\{\Phi_\Lambda(x^m)\}$ *is such that*

$$\{\mathcal{E}|L\}_{\Phi_\Lambda} \underset{v_a}{\pounds}\Phi_\Lambda = 0 \tag{4.54}$$

*holds throughout* $D^*$.

There are a number of instances in which an alternative, but equivalent formulation of the condition (4.51) is of more use. If the strong identity (4.38) is used to expand the left-hand side of (4.51), we obtain

$$\{\mathcal{E}|L\}_{\Phi_\Lambda} \underset{v_a}{\pounds}\Phi_\Lambda = \partial_m\left[\frac{\partial L}{\partial(\partial_m \Phi_\Lambda)} \underset{v_a}{D}\Phi_\Lambda\right] - W_k{}^m \partial_m v_a{}^k$$
$$+ \{\mathcal{E}|L\}_{\Phi_\Lambda} v_a{}^k \partial_k \Phi_\Lambda + v_a{}^k \partial_k^* L ,$$

Thus, we have

$$W_k{}^m \partial_m v_a{}^k = v^k \partial_k^* L + \{\mathcal{E}|L\}_{\Phi_\Lambda} \underset{v_a}{D} \Phi_\Lambda + \partial_m \left[ \frac{\partial L}{\partial(\partial_m \Phi_\Lambda)} \underset{v_a}{D} \Phi_\Lambda \right]$$

on noting that $\underset{v_a}{D} \Phi_\Lambda = v_a{}^k \partial_k \Phi_\Lambda - \underset{v_a}{\pounds} \Phi_\Lambda$. Now,

$$\partial_m \left[ \frac{\partial L}{\partial(\partial_m \Phi_\Lambda)} \underset{v_a}{D} \Phi_\Lambda \right] = \left( -\{\mathcal{E}|L\}_{\Phi_\Lambda} + \frac{\partial L}{\partial \Phi_\Lambda} \right) \underset{v_a}{D} \Phi_\Lambda + \frac{\partial L}{\partial(\partial_m \Phi_\Lambda)} \partial_m \underset{v_a}{D} \Phi_\Lambda ,$$

and hence we have the following conclusion: *the conditions expressed by* (4.51) *are equivalent to the conditions*

$$W_k{}^m \partial_m v_a{}^k = v_a{}^k \partial_k^* L + \frac{\partial L}{\partial \Phi_\Lambda} \underset{v_a}{D} \Phi_\Lambda + \frac{\partial L}{\partial(\partial_m \Phi_\Lambda)} \partial_m \underset{v_a}{D} \Phi_\Lambda . \tag{4.55}$$

Suppose that each of the $\Phi_\Lambda$ is a scalar function (that is, $F_\Lambda(\Phi_\Sigma, x^m, x^{m\prime}) = \Phi_\Lambda$ for all $\Lambda = 1, \ldots, N$), then

$$\underset{v_a}{\pounds} \Phi_\Lambda = v_a{}^k \partial_k \Phi_\Lambda ,$$

and hence

$$\underset{v_a}{D} \Phi_\Lambda = v_a{}^k \partial_k \Phi_\Lambda - \underset{v_a}{\pounds} \Phi_\Lambda = 0 .$$

This is as it should be since a scalar function by definition is unchanged under coordinate transformations. When this observation is combined with (4.55), we have the following: *if each* $\Phi_\Lambda(x)$ *is a scalar function, then* $J[\{\Phi_\Lambda\}](L; D)$ *is coordinate invariant under the action of* $G_r$ *in a sufficiently small neighborhood of the identity transformation if and only if* $\{\Phi_\Lambda(x^m)\}$ *and* $v_a^m(x^k)$ *are such that*

$$W_k{}^m \partial_m v_a{}^k = v_a{}^k \partial_k^* L . \tag{4.56}$$

Since $\{\Phi_\Lambda\}$ is assumed to be a linear geometric object field, all of the equations obtained in this and the following sections are linear in the $n$ unknown function $v^k(x^m)$. It thus follows that *if* $v_a{}^k(x^m)$, $a = 1, \ldots, r$, *are any* $r$ *solutions of the equations obtained, then*

$$*v^k(x^m) = C^a v_a{}^k(x^m) ,$$

*where* $C^a$ *are constants, is also a solution.* Also, suppose that the $r$ solutions $v_a{}^k(x^m)$ are linearly indpendent with constant coefficients. We may then ask whether the $r$ solutions form the infinitesimal generating vectors of an $r$-parameter group. For this to be the case, the theorems of Lie state that it is sufficient to test whether

$$(X_a, X_b)f = C_{ab}{}^c X_c f , \quad C_{ab}{}^c + C_{ba}{}^c = 0$$

hold when $C_{ab}{}^c$ are constants and

$$X_a = v_a{}^k(x^m)\frac{\partial}{\partial x^k} .$$

If this is the case, we are finished. If this is not the case, we must look for more solutions that are linearly independent with constant coefficients and also linearly independent with constant coefficients of the solutions already obtained and test again. What this amounts to is that if $(X_1, X_2)$ cannot be written as a linear combination of the original $X$, then $\underset{v_1}{\pounds}\, v_2{}^k(x^m) = V^k(x^m)$ must also be a solution of the system of equations obtained, since $(X_1, X_2)f = V^k(x^m)(\partial f/\partial x^k)$. This problem contains computational difficulties and can lead to rather serious questions. In fact, it may be impossible to write all solution of the system of equations as a linear combination of a finite number of linearly independent solutions. The nonfinite case is examined in Sec. 4.6.

## 4.4. WEAK POINT INVARIANCE

We shall now make inquires similar to those of Sec. 4.3 but for the case where both function variations and coordinate variations are considered and the total variation vanishes throughout $D^*$. The situation is thus considered in which

$$\underset{v}{\Delta}\Phi_\Lambda = \underset{v}{D}\Phi_\Lambda + \delta\Phi_\Lambda = 0 \tag{4.57}$$

holds throughout $D^*$, and hence we have

$$\delta\Phi_\Lambda = -\underset{v}{D}\Phi_\Lambda = \underset{v}{\pounds}\Phi_\Lambda - v^k\partial_k\Phi_\Lambda . \tag{4.58}$$

As pointed out in Sec. 4.2, this is just the case in which the manifold is subjected to a point transformation with infinitesimal generator $v^k(x^m)$. Previous calculations (4.42) give us

$$\left.\frac{dJ}{d\pi}\right|_{\pi=0} = -\int_d \left[\{\mathcal{E}|L\}_{\Phi_\Lambda} v^k\partial_k\Phi_\Lambda + \frac{\partial}{\partial x^m}(W_k{}^m v^k)\right] dv(x) . \tag{4.49}$$

The factor $v^k \partial_k \Phi_\Lambda$ which multiplies the Euler-Lagrange derivative in (4.59), shows that we are indeed in the case where a point transformation is considered since this quantity is the change per unit $\lambda$ in the field $\Phi_\Lambda$ that arises from an evaluation at the point $(x^m)$ and at the point $['x^m = x^m + d\lambda\, v^m(x^k)]$.

With the understanding that we are interested in the change in the value of $J[\{\Phi_\Lambda\}](L;D)$ for a given $\{\Phi_\Lambda\}$, as introduced in Sec. 4.3 (considerations are for fixed $\{\Phi_\Lambda(x^m)\}$, not for all $\{\Phi_\Lambda\}$ in $\mathfrak{D}_1(D^*;N)$), the following definition is given. *A functional* $J[\{\Phi_\Lambda\}](L;D)$ *is said to be weakly point invariant under a one-parameter group* $G_1$ *with infinitesimal generator* $v^k(x^m)$ *if and only if* $\underset{v}{\Delta}\Phi_\Lambda = 0$ *and* $dJ/d\pi|_{\pi=0}$ *vanishes for all* $d \subset D$. Using the reasoning of Sec. 4.3, we establish the following result.

**Theorem 4.3.** *Let L be a given lagrangian function and let* $\{\Phi_\Lambda(x^m)\}$ *be a given N-tuple of functions which constitute an element of* $\mathfrak{D}_1(D^*;N)$. *If* $J[\{\Phi_\Lambda\}](L;D)$ *is weakly point invariant under the one-parameter group*

$$G_1:\ 'x^m = G^m(x^k,\lambda) = x^m + \lambda v^m(x^k) + o(\lambda)\,, \tag{4.60}$$

*then the quantities* $\Phi_\Lambda(x^m)$ *and* $v^m(x^k)$ *must satisfy the relations*

$$\frac{\partial}{\partial x^m}(W_k{}^m v^k) = -\{\mathfrak{E}|L\}_{\Phi_\Lambda} v^k \partial_k \Phi_\Lambda\,. \tag{4.61}$$

*If* $\{\Phi_\Lambda(x^m)\}$ *and* $v^m(x^k)$ *satisfy the relations* (4.61) *then* $J[\{\Phi_\Lambda\}](L;D)$ *is weakly point invariant under the one-parameter group* $G_1$ *of point transformations restricted to a sufficiently small neighborhood of the identity transformation. Further, if* $J[\{\Phi_\Lambda\}](L;d)$ *is weakly point invariant under an* $r$*-parameter group* $G_r$ *of point transformations with the infinitesimal generators* $v_a{}^m(x^k)$, $a = 1,\ldots,r$, *then* $\{\Phi_\Lambda(x^m)\}$ *and* $v_a{}^m(x^k)$ *must satisfy the* $r$ *relations*

$$\frac{\partial}{\partial x^m}(W_k{}^m v_a{}^k) = -\{\mathfrak{E}|L\}_{\Phi_\Lambda} v_a{}^k \partial_k \Phi_\Lambda\,. \tag{4.62}$$

*If the relations* (4.62) *hold, then* $J[\{\Phi_\Lambda\}](L;D)$ *is weakly point invariant under the* $r$*-parameter group* $G_r$ *of point transformations restricted to a sufficiently small neighborhood of the identity.*

The above results can be stated in an equivalent and more useful form. If we expand the left-hand side of (4.62) and make use of the strong identity (4.38), we have

$$\partial_m (W_k{}^m v_a{}^k) = W_k{}^m \partial_m v_a{}^k + v_a{}^k \partial_m W_k{}^m$$

$$= W_k{}^m \partial_m v_a{}^k - v_a{}^k [\{\mathcal{E}|L\}_{\Phi_\Lambda} \partial_k \Phi_\Lambda + \partial_k^* L] .$$

Hence, *the conditions expressed by* (4.62) *are equivalent to the conditions*

$$W_k{}^m \partial_m v_a{}^k = v_a{}^k \partial_k^* L . \tag{4.63}$$

This result leads to the following observations. First, note that (4.63) is identical with (4.56), which results under coordinate transformations when all $N\ \Phi_\Lambda$ are scalar function on $D$. However, when point transformations are considered, the variation induced in $\Phi_\Lambda$ arises solely from evaluation at two different points and has no connection with how $\Phi_\Lambda$ transforms under coordinate transformations. It thus follows, both analytically and heuristically, that the conditions for weak point invariance can be obtained from the conditions for weak coordinate invariance by treating all $\Phi_\Lambda$ as if they were scalar functions in the results obtained for coordinate invariance.

Finally, the following alternative form results in the case in which $\{\Phi_\Lambda(x^m)\} \in S(L)$: *if* $\{\Phi_\Lambda(x^m)\} \in S(L)$, *then* $J[\{\Phi_\Lambda(x^m)](L; D)$ *is weakly point invariant in a sufficiently small neighborhood of the identity of* $G_1$ *if and only if and* $v^k(x^m)$ *are such that*

$$\partial_m (W_k{}^m v^k) = 0 . \tag{4.64}$$

In this case, $\{\Phi_\Lambda(x^m)\}$ satisfies a conservation law (4.64). If there is an $r$-parameter group $G_r$ involved, then $\{\Phi_\Lambda(x^m)\}$ satisfies the $r$ conservation laws

$$\partial_m (W_k{}^m v_a{}^k) = 0 . \tag{4.65}$$

In every instance, since $\{\Phi_\Lambda(x^m)\}$ is assumed to be given, we obtain a partial differential equation for the infinitesimal generator $v^k(x^m)$ when only $G_1$ is considered. Suppose that $v^k(x^m)$ satisfies (4.61) for given $\{\Phi_\Lambda(x^m)\}$, then

$$G^m(x^k, \lambda) = \left\{ \exp\left[ \lambda v^k(x^t) \frac{\partial}{\partial x^k} \right] \right\} x^m \tag{4.66}$$

defines a one-parameter group in its domain of convergence, and this group is such that it leaves the value of $J[\{\Phi_\Lambda\}](L; D)$ unchaged in value for the given $\{\Phi_\Lambda(x)\}$.

Now, equation (4.61) or (4.65) is only one equation in the $n$ unknown functions $v^k(x^m)$, and hence, for any particular $\{\Phi_\Lambda(x^m)\} \in S(L)$, there are many solutions. For instance, suppose that $\det(W_k{}^m) \neq 0$ throughout $D^*$; then there exists a unique collection of functions $H_k{}^m$ such that

$$W_m{}^t H_k{}^m = \delta_k{}^t .$$

These functions $H_k{}^m$ are functions of $x$ on $D^*$ since the functions $W_k{}^m$ are functions of $x$ on $D^*$ by the assumption that $\{\Phi_\Lambda(x^m)\}$ is a given, and hence a known collection of functions on $D^*$. On setting $v^k = H_t{}^k M^t$, we have

$$\partial_m(W_k{}^m v^k) = \partial_m(W_k{}^m H_t{}^k M^t) = \partial_m M^m .$$

Hence $v^k = H_t{}^k M^t$ satisfies (4.64) for any $M^m(x^k)$ such that $\partial_m M^m = 0$ (that is, $M^m = \partial_t S^{tm}$ with $S^{tm} = -S^{mt}$), in which case $v^k = H_t{}^k \partial_m S^{mt}$ for all $S^{mt} \in C^2$ with $S^{tm} = -S^{mt}$.

## 4.5. STRONG COORDINATE AND POINT INVARIANCE

In both weak coordinate invariance and weak point invariance, it was assumed that $\Phi_\Lambda(x^m)$ were given functions constituting an element of $\mathfrak{D}_1(D^*; N)$. The infinitesimal generators $v_a{}^k(x^m)$, obtained by the requirement of weak invariance, accordingly depend intrinsically on the particular functions of $x^m$ picked for $\{\Phi_\Lambda\}$. The results of the previous sections are thus not very useful for we have no way of knowing whether a set of $v_a{}^k$ that leads to a weak invariance for one set of $\{\Phi_\Lambda\}$ will lead to a weak invariance for another set of $\{\Phi_\Lambda\}$. A method of avoiding this difficulty is to look for $v_a{}^k$ that leads to weak invariance for every $\{\Phi_\Lambda(x)\}$ in $\mathfrak{D}_1(D^*; N)$. We thus give the following definitions.

*If $v^k(x^m)$ is such that $J[\{\Phi_\Lambda\}](L; D)$ is weakly coordinate invariant for every $\{\Phi_\Lambda(x^m)\}$ in $\mathfrak{D}_1(D^*; N)$, then $J[\{\Phi_\Lambda\}](L; D)$ is said to be strongly coordinate invariant and $v^k(x^m)$ is said to generate a coordinate invariance group of $J[\{\Phi_\Lambda\}](L; D)$.*

*If $v^k(x^m)$ is such that $J[\{\Phi_\Lambda\}](L; D)$ is weakly point invariant for every $\{\Phi_\Lambda(x^m)\}$ in $\mathfrak{D}_1(D^*; N)$, then $J[\{\Phi_\Lambda\}](L; D)$ is said to be strongly point invariant and $v^k(x^m)$ is said to generate a point invariance group of $J[\{\Phi_\Lambda\}](L; D)$.*

From these definitions of strong invariance, it follows immediately that the conditions for strong invariance are obtained from the conditions for weak invariance by the requirement that the conditions for weak invariance be identities over $\mathfrak{D}_1(D^*;N)$. The conclusions of Secs. 4.3 and 4.4 thus give the following results.

Theorem 4.4. *A necessary and sufficient condition that* $v^k(x^m)$ *generate a coordinate invariance group of* $J[\{\Phi_\Lambda\}](L;D)$ *is that*

$$\frac{\partial}{\partial x^m}\left\{\frac{\partial L}{\partial(\partial_m \Phi_\Lambda)} \underset{v}{D} \Phi_\Lambda - W_k{}^m v^k\right\} = \{\mathcal{E}|L\}_{\Phi_\Lambda} \underset{v}{\pounds} \Phi_\Lambda \tag{4.67}$$

*or*

$$W_k{}^m \partial_m v^k = v^k \partial_k^* L + \frac{\partial L}{\partial \Phi_\Lambda} \underset{v}{D} \Phi_\Lambda + \frac{\partial L}{\partial(\partial_m \Phi_\Lambda)} \partial_m \underset{v}{D} \Phi_\Lambda \tag{4.68}$$

*be identities over* $\mathfrak{D}_1(D^*;N)$. *A necessary and sufficient condition that* $v^k(x^m)$ *generate a point invariance group of* $J[\{\Phi_\Lambda\}](L;D)$ *is that*

$$\frac{\partial}{\partial x^m}(W_k{}^m v^k) = -\{\mathcal{E}|L\}_{\Phi_\Lambda} v^k \partial_k \Phi_\Lambda \tag{4.69}$$

*or*

$$W_k{}^m \partial_m v^k = v^k \partial_k^* L \tag{4.70}$$

*be identities over* $\mathfrak{D}_1(D^*;N)$.

Obviously, if there are $r$ independent $v_a{}^k(x^m)$, each of which generates a coordinate invariance (point invariance) group, and if

$$\begin{aligned}
&\text{i.}\quad C^a v_a{}^k(x^m) = 0\,, \quad \partial_m C^a = 0 \Longrightarrow C^a = 0\\
&\text{ii.}\quad (\underset{v_a}{\pounds}\,\underset{v_b}{\pounds} - \underset{v_b}{\pounds}\,\underset{v_a}{\pounds})\Phi_\Lambda = C_{ab}{}^c \underset{c}{\pounds}\Phi_\Lambda\\
&\text{iii.}\quad C_{ab}{}^c = -C_{ba}{}^c\,, \quad \partial_m C_{ab}{}^c = 0\,,
\end{aligned} \tag{4.71}$$

then the $r$ vectors $v_a{}^k(x^m)$ are the infinitesimal generators of an $r$-parameter coordinate invariance (point invariance) group. This is the second part of the second fundamental theorem of Lie on continuous groups (see, for instance, Schouten, J. A., *Ricci Calculus*, Springer, Berlin (1954)). The explicit statement, in the case of $r$-parameter invariance groups is so simple, in view of Theorem 4.4 and the conditions (4.71), that we shall not state them explicitly.

As immediate consequences, however, we have the following results.

**Theorem 4.5.** *If* $v_a^k(x^m)$ *generate an* $r$*-parameter coordinate invariance group of* $J[\{\Phi_\Lambda\}](L;D)$, *then every element of* $S(L)$ *satisfies the* $r$ *conservation laws*

$$\partial_m \left\{ \frac{\partial L}{\partial(\partial_m \Phi_\Lambda)} \underset{v_a}{D} \Phi_\Lambda - W_k{}^m v_a{}^k \right\} = 0, \quad a = 1, \ldots, r. \tag{4.71}$$

*If* $v_a^k(x)$ *generate an* $r$*-parameter point invariance group of* $J[\{\Phi_\Lambda\}](L;D)$, *then every element of* $S(L)$ *satisfies the* $r$ *conservation laws*

$$\partial_m (W_k{}^m v_a{}^k) = 0, \quad a = 1, \ldots, r. \tag{4.73}$$

We thus see the very intimate relation between conservation laws admitted of a system of Euler equations $[\{\Phi_\Lambda\} \in S(L)]$ and the invariance transformations of the functional that gives rise to the Euler equations. Historically, it is this relation which actually led Lie and others to consider continuous $r$-parameter groups, although the motivation for their considerations came from the calculus of variations in only one indpendent variable and the relation was between invariance of the classical action and quadratures of a dynamical system.

It should now be evident that strong invariance is what is usually of interest,; weak invariance is usually used only if we are interested in certain properties of a particular $\{\Phi_\Lambda(x^m)\}$ which is presented to us as being of interest from some external consideration. Unfortunately, it is a rather difficult task, except in the simplest of cases, to obtain the generators of an invariance transformation from Theorem 4.5. Let us look at just what is required. We have to be given the function space $\mathfrak{D}_1(D^*;N)$, and we have to know the geometry class to which the elements of this function space belong under action by $G_n^u$. We have to know whether we are dealing with collections of scalars or collections of vectors or whatever. We then have to be given the lagrangian function, and finally we have to demand that conditions such as (4.67), (4.68), (4.69), or (4.70) are identically satisfied over $\mathfrak{D}_1(D^*;N)$. This identical satisfaction demands that the coefficients of $\Phi_\Lambda(x^m)$ and of its derivatives each vanish separately. Accordingly, we obtain many equations which $v^k(x^m)$ must satisfy. In general this gives a rather formidable array of simultaneous first-order partial differential equations which we must solve and then use the conditions (4.71) to select the generators of the group being sought—all in all a lengthy and thankless task. It is thankless for the simple reason that we usually know $v_a{}^k(x^m)$ in advance, since the physics tells us the invariances of the action functional $J[\{\Phi_\Lambda\}](L;D)$. What is usually not known in advance is the functional form of the lagrangian function since the physics does not give us this function for the asking. The real problem is again one step removed from the present considerations and will be considered in detail later. For the present, let us examine a particularly simple case in which the calculations are straightforward.

Let $\{\Phi_\Lambda(x^m)\}$ consist of the single scalar function $\varphi(x)$, so that $N = 1$, and let $x^1 = x$, $x^2 = t$ with $n = 2$. As a lagrangian function, we take

$$L = \tfrac{1}{2}[(\partial_x \varphi)^2 - a^2(\partial_t \varphi)^2 - b^2 \varphi^2] \tag{4.74}$$

in which $a$ and $b$ are given, nonzero numbers. In this case, we have as the Euler equation,

$$\{\mathcal{E}|L\}\varphi = -b^2\varphi - \partial_x \partial_x \varphi + a^2 \partial_t \partial_t \varphi = 0, \tag{4.75}$$

i.e., the "one-dimensional Klein-Gordon equation." Since $\varphi$ is a scalar function, $\underset{v}{D}\varphi = 0$. Thus, both (4.68) and (4.70) give

$$W_k{}^m \partial_m v^k = v^k \partial_k^* L = 0, \tag{4.76}$$

since $\partial_k^* L = 0$ by (4.74). A straightforward calculation yields

$$W_x{}^x = \tfrac{1}{2}[(\partial_x\varphi)^2 + a^2(\partial_t\varphi)^2 + b^2\varphi^2], \quad W_t{}^x = \partial_t\varphi\,\partial_x\varphi,$$

$$W_x{}^t = -a^2 W_t{}^x, \quad W_t{}^t = -\tfrac{1}{2}[(\partial_x\varphi)^2 + a^2(\partial_t\varphi)^2 - b^2\varphi^2],$$

and hence (4.76) becomes

$$\tfrac{1}{2}[(\partial_x\varphi)^2 + a^2(\partial_t\varphi)^2](\partial_x v^x - \partial_t v^t) + \partial_t\varphi\,\partial_x\varphi\,[\partial_x v^t - a^2\partial_t v^x] + \tfrac{1}{2}b^2\varphi^2[\partial_x v^x + \partial_t v^t] = 0. \tag{4.77}$$

By the above theorems, $(v^x, v^t)$ will generate an invariance transformation (both point and coordinate in this case since $\varphi$ is scalar) if and only if (4.77) is an identity over $\mathfrak{D}_1(D^*;1)$. This can be the case only if

$$\partial_x v^x - \partial_t v^t = 0\,, \tag{4.78}$$

$$\partial_x v^t - a^2 \partial_t v^x = 0\,, \tag{4.79}$$

$$\partial_x v^x + \partial_t v^t = 0\,, \tag{4.80}$$

since $a \neq 0$, $b \neq 0$. Now, (4.78) and (4.80) give $\partial_x v^x = \partial_t v^t = 0$, and we have the system

$$\partial_x v^t - a^2 \partial_t v^x = \partial_x v^x = \partial_t v^t = 0\,. \tag{4.81}$$

All solutions of the system (4.81) can be expressed as linear combinations of the three solutions

$$\begin{matrix} v_1{}^x = 1 \\ v_1{}^t = 0 \end{matrix}\,,\quad \begin{matrix} v_2{}^x = 0 \\ v_2{}^t = 1 \end{matrix}\,,\quad \begin{matrix} v_3{}^x = t \\ v_3{}^t = a^2 x\,. \end{matrix} \tag{4.82}$$

These three solutions satisfy the conditions (4.71). We thus have a three-parameter invariance group for $J[\varphi](L;D)$. The finite groups generated by these infinitesimal generators are obtained by

$$G_a{}^m(x, t, \lambda) = \exp[v_a{}^k \partial_k] x^m\,.$$

Thus, we have

$$\begin{cases} v_1{}^k \;\Rightarrow \begin{cases} x' = x + \lambda \\ t' = t \end{cases} \\ v_2{}^k \;\Rightarrow \begin{cases} x' = x \\ t' = t + \lambda \end{cases} \\ v_3{}^k \;\Rightarrow \begin{cases} x' = x \cosh(a\lambda) + (t/a)\sinh(a\lambda) \\ t' = ax\sinh(a\lambda) + t\cosh(a\lambda)\,, \end{cases} \end{cases} \tag{4.83}$$

and the three-parameter invariance group of $J[\varphi](L;D)$ is the three-parameter Lorentz group on the two dimensional space $(x, t)$.

So far, we have assumed that the lagrangian function is given; and, indeed, an inspection of the results obtained above shows just how intimately strong invariance is tied to the particular lagrangian function under consideration. In practice, however, the lagrangian function is not given and must be obtained starting from a system of partial differential equations which are identified with the Euler equations that arise from the lagrangian function which we require. Obtained in this way, the lagrangian function is not unique—it is determined only to within an equivalence class: if $L$ is such a lagrangian function, then all elements of the equivalence class $\hat{L}$ (comprised of elements $L + A$ for all $A \in \mathfrak{N}(N)$) have the same Euler equations. We have seen that $W_k{}^m$ is linear with respect to the lagrangian function,

$$W_k{}^m(L + A) = W_k{}^m(L) + W_k{}^m(A) . \tag{4.84}$$

Regarding the Euler equations, we cannot distinguish between $L$ and $L + A$ for any $A \in \mathfrak{N}(N)$. It thus follows that there will be many invariance groups associated with a given system of Euler equations. A more precise statement is as follows.

Theorem 4.6. *Let* L *be a lagrangian function which embeds a given system of equations in a variational statement on a space* $\mathfrak{D}_1(D^*; N)$, *and let* $\mathfrak{N}(N)$ *be the null class of the Euler-Lagrange operator on* $\mathfrak{D}_1(D^*; N)$. *For each element* A *of* $\mathfrak{N}(N)$ *we have a coordinate invariance group defined by the requirement that*

$$\partial_m \left\{ \frac{\partial(L + A)}{\partial(\partial_m \Phi_\Lambda)} \underset{v}{D} \Phi_\Lambda - W_k{}^m(L + A)\, v^k \right\} = \{\mathcal{E}|L + A\}_{\Phi_\Lambda} \underset{v}{\mathcal{L}} \Phi_\Lambda \tag{4.85}$$

*or*

$$W_k{}^m(L + A)\, \partial_m v^k = v^k \partial_k^*(L + A) + \frac{\partial(L + A)}{\partial \Phi_\Lambda} \underset{v}{D} \Phi_\Lambda + \frac{\partial(L + A)}{\partial(\partial_m \Phi_\Lambda)} \partial_m \underset{v}{D} \Phi_\Lambda \tag{4.86}$$

*be an identity over* $\mathfrak{D}_1(D^*; N)$ *and a point invariant group defined by the requirement that*

$$\partial_m \{W_k{}^m(L + A)\, v^k\} = -\{\mathcal{E}|L + A\}_{\Phi_\Lambda} v^k \partial_k \Phi_\Lambda \tag{4.87}$$

*or*

$$W_k{}^m(L + A)\, \partial_m v^k = v^k \partial_k^*(L + A) \tag{4.88}$$

*be an identity over* $\mathfrak{D}_1(D^*; N)$.

It is evident from Theorem 4.6 that the number of parameters in an invariance group will depend on which $A$ is chosen from $\mathfrak{N}(N)$.

One way out of this problem of many different invariance groups being associated with a given system of Euler equations is to pick the element of $\mathfrak{N}(N)$ for which the invariance group is of maximal order. Even in this case, there is no full resolution, for there may be several inequivalent invariance groups of the same order that can be obtained by different choices of elements of $\mathfrak{N}(N)$. The full resolution of this problem is an open question, and attention must be directed back to the physics and what is actually known in the physical context. From this point of view, we know certain symmetry properties which the fields must satisfy—we know the generators of a group which must be an invariance group for the action—but we do not know the lagrangian function itself. What is thus required is to seek the set of all lagrangian functions for which the given group is an invariance group, either coordinatewise or pointwise. First, however, certain other types of invariance are considered.

To observe what actually occurs, let us look at the example considered previously. From the result of Chap. 1, we know that

$$A = K^k(\partial_k\varphi)\alpha\varphi^{\alpha-1}, \quad \partial_m K^k = 0, \quad \alpha > 0 \tag{4.89}$$

is an element of $\mathfrak{N}(1)$. With $L$ given by (4.74), $L + A$ leads to the same Euler equations (4.75). Since $\partial_k^* A = 0$, $\partial_k^*(L + A) = 0$, we have the condition that

$$[W_k{}^m(L) + W_k{}^m(A)]\,\partial_m v^k = 0 \tag{4.90}$$

shall be an identity on $\mathfrak{D}_1(D^*; 1)$. In expanded form, this is equivalent to

$$\begin{aligned}
&\tfrac{1}{2}[(\partial_x\varphi)^2 + a^2(\partial_t\varphi)^2](\partial_x v^x - \partial_t v^t) + \partial_t\varphi\,\partial_x\varphi\,[\partial_x v^t - a^2\partial_t v^x] \\
&+ \tfrac{1}{2}b^2\varphi^2[\partial_x v^x + \partial_t v^t] + \alpha\varphi^{\alpha-1}\partial_t\varphi\,[K^x\partial_x v^t - K^t\partial_x v^x] \\
&+ \alpha\varphi^{\alpha-1}\partial_x\varphi\,[K^t\partial_t v^x - K^x\partial_t v^t] = 0 .
\end{aligned} \tag{4.91}$$

In order that this identity hold, we must now have

$$\partial_x v^x - \partial_t v^t = \partial_x v^x + \partial_t v^t = \partial_x v^t - a^2\partial_t v^x = 0 ,$$

as obtained previously, and in addition

$$K^x\partial_x v^t - K^t\partial_x v^x = K^t\partial_t v^x - K^x\partial_t v^t = 0 .$$

Hence, if either $K^x \neq 0$ or $K^t \neq 0$ (that is $A \neq 0$) we must have $v^x =$ constant, $v^t =$ constant, and, accordingly, only the two-parameter invariance group

$$\begin{cases} v_1^{\ x} = 1\,, & x' = x + \lambda \\ v_1^{\ t} = 0\,, & t' = t \end{cases}$$
$$\begin{cases} v_2^{\ x} = 0\,, & x' = x \\ v_2^{\ t} = 1\,, & t' = t + \lambda\,. \end{cases} \tag{4.92}$$

The group (4.92) is a proper subgroup of the three-parameter Lorentz group (4.83). In this case the invariance group for $L + A$ is different from the invariance group for $L$, even for such a simple $A$, and is in fact a subgroup of that obtained for $L$.

## 4.6. ABSOLUTE INVARIANCE

Up to this point, our considerations of invariance have been with respect to transformations which can be represented in terms of a finite number of infinitesimal generators ($r$-parameter continuous groups with finite $r$). Suppose we wish to consider invariance for more general groups, such as the group of all locally invertible transformations of class $C^u$. Such groups do not form finite parameter groups. We can analyze the effects of such groups, however, by studying the elements of the group which are in a sufficiently small neighborhood of the identity element. For such elements, we write

$$x^{m\prime} = x^m + \lambda V^m(x^k) + o(\lambda) \tag{4.93}$$

in which the $n$ functions $V^m(x^k)$ must now be allowed to range over the space of all $u$-times continuously differentiable functions on $D^*$. With this at our disposal, we give the following definitions.

*A functional* $J[\{\Phi_\Lambda\}](L;D)$ *is said to be absolutely coordinate invariant with respect to a nonfinite continuous group of coordinate transformations* (4.93) *if and only if* $dJ[\{\Phi_\Lambda\}](L;d)/d\lambda|_{\lambda=\eta=0} = 0$ *for all* $V^m(x^k) \in C^u$ *and all* $d \subset D$.

*A functional* $J[\{\Phi_\Lambda\}](L;D)$ *is said to be absolutely point invariant with respect to a nonfinite continuous group of coordinate transformations*.

$${}'x^m = x^m + \lambda V^m(x^k) + o(\lambda) \tag{4.94}$$

*if and only if* $dJ[\{\Phi_\Lambda\}](L;d)/d\pi|_{\pi=0} = 0$ *for all* $V^m(x^k)$, *all* $d \subset D$, *and* $\underset{v}{\Delta}\Phi_\Lambda = 0$.

The results established in the previous sections lead directly to the following conclusions

Theorem 4.7. *A necessary and sufficient condition that* $J[\{\Phi_\Lambda\}](L;D)$ *be absolutely coordinate invariant is that*

$$\partial_m\left\{\frac{\partial L}{\partial(\partial_m\Phi_\Lambda)}\underset{v}{D}\Phi_\Lambda - W_k{}^m V^k\right\} = \{\mathcal{E}|L\}_{\Phi_\Lambda}\underset{v}{\pounds}\Phi_\Lambda \tag{4.95}$$

*or*

$$W_k{}^m\partial_m V^k = V^k\partial_k^* L + \frac{\partial L}{\partial\Phi_\Lambda}\underset{v}{D}\Phi_\Lambda + \frac{\partial L}{\partial(\partial_m\Phi_\Lambda)}\partial_m\underset{v}{D}\Phi_\Lambda \tag{4.96}$$

*be identically satisfied for all* $V^m(x^k) \in C^u$. *A necessary and sufficient condition that* $J[\{\Phi_\Lambda\}](L;D)$ *be absolutely point invariant is that*

$$\partial_m(W_k{}^m V^k) = -\{\mathcal{E}|L\}_{\Phi_\Lambda}V^k\partial_k\Phi_\Lambda \tag{4.97}$$

*or*

$$W_k{}^m\partial_m V^k = V^k\partial_k^* L \tag{4.98}$$

*be identically satisfied for all* $V^m(x^k) \in C^u$.

Let us start with the case of absolute point invariance, since it is reasonably straightforward. By Theorem 4.7, (4.98) must be an identity for all $V^m(x^k) \in C^u$. This can be the case if and only if

$$\partial_k^* L \equiv 0\,, \quad W_k{}^m \equiv 0\,, \tag{4.99}$$

since the coefficients of $V^k$ and its derivatives must vanish seaparately. *Necessary and sufficient conditions that* $J[\{\Phi_\Lambda\}](L;D)$ *be absolutely point invariant is that the lagrangian function* L *be such that*

$$\partial_k^* L \equiv 0\,, \quad \partial_k\Phi_\Lambda\frac{\partial L}{\partial(\partial_m\Phi_\Lambda)} - \delta_k{}^m L \equiv 0\,. \tag{4.100}$$

Absolute point invariance thus requires that the lagrangian function be functionally independent of $x^m$ and that the momentum-energy complex vanish throughout $D^*$. As expected, the situation proves uninteresting, since absolute point invariance is equivalent to the statement $J[\{\Phi\}](L;D)$ is invariant in value for all $\{\Phi\} \in \mathfrak{D}_1(D^*;N)$ and for all $C^u$ methods of moving the points of $D^*$ around in $X_n{}^u$. For the case $N = 1$, we must have

$$0 \equiv W_k{}^m = \partial_k \varphi \frac{\partial L}{\partial(\partial_m \varphi)} - \delta_k{}^m L\,, \quad \partial_k^* L \equiv 0\,,$$

and hence $L = 0$.

We now consider the case of absolute coordinate invariance and work with (4.96). From the definition of $\underset{v}{D}$, we have

$$\underset{v}{D}\Phi_\Lambda = V^k \partial_k \Phi_\Lambda - \underset{v}{\pounds}\Phi_\Lambda\,. \tag{4.101}$$

The results of Chap. 3 give

$$\underset{v}{\pounds}\Phi_\Lambda = V^k \partial_k \Phi_\Lambda - \{F,V\}_\Lambda{}^\Sigma \Phi_\Sigma + \{G,V\}_\Lambda \tag{4.102}$$

for *linear* geometric objects, with the transformation of class $p$

$$\Phi_{\Lambda'} = F_\Lambda(x^m, x^{m'})\Phi_\Sigma + G_\Lambda(x, x^{m'})\,,$$

and

$$\left\{\begin{aligned} \{F,V\}_\Lambda{}^\Sigma &= \sum_{s=1}^{p} F_k{}_\Lambda^{m(s)\Sigma} \partial_{m_1} \cdots \partial_{m_s} V^k + F_{k\Lambda}{}^{m_0\Sigma} V^k \\ \{G,V\}_\Lambda &= \sum_{s=1}^{p} G_k{}^{m(s)} \partial_{m_1} \cdots \partial_{m_s} V^k + G_{k\Lambda}{}^{m_0} V^k \end{aligned}\right. \tag{4.103}$$

In these formulas, $F_{k\Lambda}{}^{m(s)\Sigma}$, $G_{k\Lambda}{}^{m(s)}$ are functions of the $x^m$ only and are determined uniquely by the functions

$$F_\Lambda{}^\Sigma(x^m, x^{m'}) \quad \text{and} \quad G_\Lambda(x^m, x^{m'})$$

which define the transformation properties of $\Phi_\Lambda$. When (4.101) to (4.103) are combined, we accordingly have

$$\begin{aligned} \underset{v}{D}\Phi_{\Lambda'} = {} & (F_{k\Lambda}{}^{m_0\Sigma}\Phi_\Lambda + G_{k\Lambda}{}^{m_0}) V^k \\ & + \sum_{s=1}^{p} (F_{k\Lambda}{}^{m(s)\Sigma}\Phi_\Sigma + G_{k\Lambda}{}^{m(s)}) \partial_{m_1} \cdots \partial_{m_s} V^k\,. \end{aligned} \tag{4.104}$$

Hence, we obtain

$$\partial_m \underset{v}{D} \Phi_\Lambda = V^k \partial_m (F_{k\Lambda}{}^{m_0\Sigma}\Phi_\Sigma + G_{k\Lambda}{}^{m_0}) + (F_{k\Lambda}{}^{m_0\Sigma}\Phi_\Sigma + G_{k\Lambda}{}^{m_0})\partial_m V^k$$
$$+ \sum_{s=1}^{p} (F_{k\Lambda}{}^{m(s)\Sigma}\Phi_\Sigma + G_{k\Lambda}{}^{m(s)})\partial_m \partial_{m_1} \cdots \partial_{m_s} V^k \tag{4.105}$$
$$+ \sum_{s=1}^{p} \partial_m (F_{k\Lambda}{}^{m(s)\Sigma}\Phi_\Sigma + G_{k\Lambda}{}^{m(s)})\partial_{m_1} \cdots \partial_{m_s} V^k .$$

When the various elements are collected together and substituted into (4.96), with care to collect similar terms in $V^k$ and their derivatives together, we have

$$0 \equiv V^k \Bigg[ \partial_k^* L + \frac{\partial L}{\partial \Phi_\Lambda}(F_{k\Lambda}{}^{m_0\Sigma}\Phi_\Sigma + G_{k\Lambda}{}^{m_0})$$
$$+ \frac{\partial L}{\partial(\partial_m \Phi_\Lambda)} \partial_m (F_{k\Lambda}{}^{m_0\Sigma}\Phi_\Sigma + G_{k\Lambda}{}^{m_0}) \Bigg]$$
$$+ \partial_t V^k \Bigg[ -W_k{}^t + \frac{\partial L}{\partial \Phi_\Lambda}(F_{k\Lambda}{}^{t\Sigma}\Phi_\Sigma + G_{k\Lambda}{}^t)$$
$$+ \frac{\partial L}{\partial(\partial_t \Phi_\Lambda)}(F_{k\Lambda}{}^{m_0\Sigma}\Phi_\Sigma + G_{k\Lambda}{}^{m_0})$$
$$+ \partial_m (F_{k\Lambda}{}^{t\Sigma}\Phi_\Sigma + G_{k\Lambda}{}^t) \frac{\partial L}{\partial(\partial_m \Phi_\Lambda)} \Bigg]$$
$$+ \partial_{t_1}\partial_{t_2} V^k \Bigg[ \frac{\partial L}{\partial \Phi_\Lambda}(F_{k\Lambda}{}^{t_1 t_2 \Sigma}\Phi_\Sigma + G_k{}^{t_1 t_2})$$
$$+ \frac{\partial L}{\partial(\partial_m \Phi_\Lambda)} \partial_m (F_{k\Lambda}{}^{t_1 t_2 \Sigma}\Phi_\Sigma + G_{k\Lambda}{}^{t_1 t_2})$$
$$+ \frac{\partial L}{\partial(\partial_{t_1} \Phi_\Lambda)}(F_{k\Lambda}{}^{t_2\Sigma}\Phi_\Sigma + G_{k\Lambda}{}^{t_2}) \Bigg]$$
$$+ \partial_{t_1}\partial_{t_2}\partial_{t_3} V^k \Bigg[ \frac{\partial L}{\partial \Phi_\Lambda}(F_{k\Lambda}{}^{t_1 t_2 t_3 \Sigma}\Phi_\Sigma + G_{k\Lambda}{}^{t_1 t_2})$$
$$+ \frac{\partial L}{\partial(\partial_m \Phi_\Lambda)} \partial_m (F_{k\Lambda}{}^{t_1 t_2 t_3 \Sigma}\Phi_\Sigma + G_{k\Lambda}{}^{t_1 t_2 t_3}) +$$

$$+ \frac{\partial L}{\partial(\partial_{t_1}\Phi_\Lambda)}(F_{k\Lambda}{}^{t_2 t_3 \Sigma}\Phi_\Sigma + G_{k\Lambda}{}^{t_2 t_3})\Big]$$

$$+ \cdots + \partial_{t_1} \cdots \partial_{t_{p+1}} V^k \frac{\partial L}{\partial(\partial_{t_1}\Phi_\Lambda)}(F_{k\Lambda}{}^{t_2 \cdots t_p \Sigma}\Phi_\Sigma + G_{k\Lambda}{}^{t_2 \cdots t_p}) \,. \tag{4.106}$$

If this is to be identically satisfied for all $V^k \in C^u$, the coefficients of $V^k$ and its derivatives must vanish separately. However, $V^k \in C^u$, $u > p + 2$ and hence $\partial_{t_1}\partial_{t_2}V^k \equiv \partial_{t_2}\partial_{t_1}V^k$ Hence, if there is more than one $t$, only the symmetric combinations in $t$ must vanish. If we denote symmetrization with respect to $t$ by $S(t)$, we have established the following results.

**Theorem 4.8.** *If* $\{\Phi_\Lambda(x)\}$ *is a linear geometric object of class* $p$, *then* $J[\{\Phi_\Lambda\}](L; D)$ *is absolutely coordinate invariant if and only if* $L$ *is such that the following identities hold on* $\mathfrak{D}_1(D^*; N)$:

$$0 = \partial_k^* L + \frac{\partial L}{\partial \Phi_\Lambda}(F_{k\Lambda}{}^{m_0 \Sigma}\Phi_\Sigma + G_{k\Lambda}{}^{m_0})$$
$$+ \frac{\partial L}{\partial(\partial_m \Phi_\Lambda)}\partial_m(F_{k\Lambda}{}^{m_0 \Sigma}\Phi_\Sigma + G_{k\Lambda}{}^{m_0}) \,, \tag{4.107}$$

$$0 = -W_k{}^t + \frac{\partial L}{\partial \Phi_\Lambda}(F_{k\Lambda}{}^{t\Sigma}\Phi_\Sigma + G_{k\Lambda}{}^t) + \frac{\partial L}{\partial(\partial_t \Phi_\Lambda)}(F_{k\Lambda}{}^{m_0 \Sigma}\Phi_\Sigma + G_{k\Lambda}{}^{m_0}) \tag{4.108}$$
$$+ \frac{\partial L}{\partial(\partial_m \Phi_\Lambda)}\partial_m(F_{k\Lambda}{}^{t\Sigma}\Phi_\Sigma + G_{k\Lambda}{}^t) \,,$$

$$0 = S(t)\Big[\frac{\partial L}{\partial \Phi_\Lambda}(F_{k\Lambda}{}^{t_1 t_2 \Sigma}\Phi_\Sigma + G_{k\Lambda}{}^{t_1 t_2})$$
$$+ \frac{\partial L}{\partial(\partial_m \Phi_\Lambda)}\partial_m(F_{k\Lambda}{}^{t_1 t_2 \Sigma}\Phi_\Sigma + G_{k\Lambda}{}^{t_1 t_2}) \tag{4.109}$$
$$+ \frac{\partial L}{\partial(\partial_{t_1}\Phi_\Lambda)}(F_{k\Lambda}{}^{t_2 \Sigma}\Phi_\Sigma + G_{k\Lambda}{}^{t_2})\Big] \,,$$

$$0 = S(t)\Big[\frac{\partial L}{\partial \Phi_\Lambda}(F_{k\Lambda}{}^{t_1 \cdots t_r \Sigma}\Phi_\Sigma + G_{k\Lambda}{}^{t_1 \cdots t_r})$$
$$+ \frac{\partial L}{\partial(\partial_m \Phi_\Lambda)}\partial_m(F_{k\Lambda}{}^{t_1 \cdots t_r \Sigma}\Phi_\Sigma + G_{k\Lambda}{}^{t_1 \cdots t_r}) +$$

$$+ \frac{\partial L}{\partial(\partial_{t_1}\Phi_\Lambda)}(F_{k\Lambda}{}^{t_2\ldots t_r\Sigma}\Phi_\Sigma + G_{k\Lambda}{}^{t_2\ldots t_r})\Bigg] \tag{4.110}$$

*for* $2 < r \leq p$, *and*

$$0 = S(t)\left[\frac{\partial L}{\partial(\partial_{t_1}\Phi_\Lambda)}(F_{k\Lambda}{}^{t_2\ldots t_p\Sigma}\Phi_\Sigma + G_{k\Lambda}{}^{t_2\ldots t_p})\right]. \tag{4.111}$$

These identities are of great importance in geometrizable field theory, and in particular, in general relativity and unified field theory.

The usual case encountered in the literature is where the lagrangian function is assumed to transform under coordinate transformations as a scalar density: $L' = (x'/x)^{-1}L$, where $(x'/x)$ denotes the Jacobian of the transformation. The results obtained above make no assumptions as to how $L$ transforms, and hence includes the previous analyses as special cases. Indeed, one has no right, on an *a priori* basis, to say how the lagrangian function transforms under coordinate transformations, since it is not a primitive quantity—the Euler equations are the primitive quantities, or possibly energy considerations. If energy considerations are taken as the primitive quantities, the energy is computed only for the actual field, not for all elements of $\mathfrak{D}_1(D^*; N)$. Thus, again we have no basis of saying what must be the transformation properties of the lagrangian function.

Prior to Sec. 4.6, we have assumed that the lagrangian function is given and that we are given either an element of $\mathfrak{D}_1(D^*; N)$ or all elements of $\mathfrak{D}_1(D; N)$. We then obtained a system of partial differential equations, the solutions of which determined the generating vectors of groups yielding certain invariance properties. In the present instance, *the system of equations* (4.107) *through* (4.111) *is a system for the determination of the lagrangian function* L *for a given* $\mathfrak{D}_1(D^*; N)$. The set of all solutions of this system determines all lagrangian functions on $\mathfrak{D}_1(D^*; N)$ which lead to functionals that are absolutely coordinate invariant. Particular note should be made that the system (4.107) through (4.111) is a linear system of partial differential equations.

## 4.7. FUNCTION INVARIANCE

We now turn to a completely different kind of invariance. Let $\mathfrak{K}(r)$ denote a linear space of $r$-tuples of function $\{\psi_r(x^m)\}$ which are defined on $D$, and let $\{\psi_r{}^o\}$ denote the $r$-tuple $(0, 0, \ldots 0)$ on $D$. We consider a system of $rN$ operators $\mathcal{O}_\Lambda{}^r$ and an appropriate structure on $\mathcal{K}(r)$ such that $\{\mathcal{O}_\Lambda{}^r \psi_r\}$ is an element of $\mathfrak{D}_1(D^*; N)$ for every $\{\psi_r(x^m)\} \in \mathcal{K}(r)$. We then define the $N$ functions $\widetilde{\Phi}_\Lambda(x^m)$ by the relations

$$\widetilde{\Phi}_\Lambda(x^m) = H_\Lambda[\Phi_\Sigma; \mathcal{O}_\Sigma{}^n \psi_r(x^m)] , \tag{4.112}$$

in which the functions $H_\Lambda(\,;\,)$ are such that

i. $\{\widetilde{\Phi}_\Lambda\}$ belongs to $\mathfrak{D}_1(D^*; N)$ whenever $\{\Phi_\Lambda\}$ belongs to $\mathfrak{D}_1(D^*; N)$ and $\{\psi_r\}$ belongs to $\mathcal{K}(r)$,

ii. $H_\Lambda(\Phi_\Sigma; \mathcal{O}_\Sigma{}^r \psi_r{}^o) = \Phi_\Sigma(x^m)$ ,

iii. for any $r$-tuple of real numbers $\{\eta^r\}$, and for all $\{\psi_r\} \in \mathcal{K}(r)$

$$\begin{aligned}\widetilde{\Phi}_\Lambda(x^m; \eta^r) &= H_\Lambda(\Phi_\Sigma; \mathcal{O}_\Sigma{}^r \eta^r \psi_r) \\ &= \Phi_\Lambda + \eta^s \left.\frac{\partial H_\Lambda}{\partial(\mathcal{O}_\Sigma{}^r \lambda^r \psi_r)}\right|_{\lambda^r=0} \left.\frac{\partial(\mathcal{O}_\Sigma{}^r \lambda^r \psi_r)}{\partial \lambda^s}\right|_{\lambda^r=0} + o(\|\eta\|)\end{aligned} \tag{4.114}$$

in which $o(\|\eta\|)$ is with respect to the norm on $\mathfrak{D}_1(D^*; N)$,

iv.
$$\frac{\partial}{\partial \eta^s}\frac{\partial}{\partial x^m}\widetilde{\Phi}_\Lambda(x^m; \eta^r) = \frac{\partial}{\partial x^m}\frac{\partial}{\partial \eta^s}\widetilde{\Phi}_\Lambda(x^m; \eta^r) . \tag{4.115}$$

In most applications, the functions $H_\Lambda(\,;\,)$ have the additional property

$$H_\Lambda[H_\Sigma(\Phi_\Gamma; \mathcal{O}_\Gamma{}^r \psi_r{}^1); \mathcal{O}_\Sigma{}^r \psi_r{}^2] = H_\Lambda[\Phi_\Sigma; \mathcal{O}_\Sigma{}^r(\psi_r{}^1 + \psi_r{}^2)] \tag{4.116}$$

for all $(\{\psi_r{}^1\}, \{\psi_r{}^2\}) \in \mathcal{K}(r)$. If this latter condition is satisfied, the functions defined by **(4.112)** constitute ***an additive function group.***

By conditions **(4.113)** and **(4.114)**, $\{\widetilde{\Phi}_\Lambda\}$ is close to $\{\Phi_\Lambda\}$ in the $\mathfrak{D}_1(D^*; N)$ norm for $\|\eta\|$ sufficiently small. Accordingly, we may define the quantities $\widetilde{\delta}_s \Phi_\Lambda$ by

$$\widetilde{\delta}_s \Phi_\Lambda = \lim_{\eta^s \to 0}\left[\left.\left(\frac{\widetilde{\Phi}_\Lambda - \Phi_\Lambda}{\eta^s}\right)\right|_{\eta^1 = \ldots \eta^{s-1} = \eta^{s+1} = \ldots = \eta^r = 0}\right]$$

and obtain

$$\widetilde{\delta}_s \Phi_\Lambda = \left.\frac{\partial H_\Lambda}{\partial(\mathcal{O}_\Sigma{}^r \lambda^r \psi_r)}\right|_{\lambda^r=0} \left.\frac{\partial(\mathcal{O}_\Sigma{}^r \lambda^r \psi_r)}{\partial \lambda^r}\right|_{\lambda^r=0} \tag{4.117}$$

As examples, if $r = N = 1$, $\tilde{\Phi}_1 = e^{\psi}\Phi_1$, then $\mathcal{O}_1{}^1\psi_1 = e^{\psi}$, $H_1(\Phi, \mathcal{O}_1{}^1\psi_1) = \Phi_1\mathcal{O}_1{}^1\psi_1$ and hence $\tilde{\delta}_1\Phi_1 = \Phi_1\psi_1$. If $r = 1$, $N = n$, $\tilde{\Phi}_m = \Phi_m + \partial_m\psi_1$, then $\mathcal{O}_m{}^1\psi_1 = \partial_m\psi_1$, $H_k(\Phi_k; \mathcal{O}_k{}^1\psi_1) = \Phi_k + \mathcal{O}_k{}^1\psi_1$, and hence $\tilde{\delta}_1\Phi_m = \partial_m\psi$ In both cases $\tilde{\Phi}$ forms an additive function group.

Since $\{\tilde{\Phi}_\Lambda\}$ is required to be an element of $\mathfrak{D}_1(D^*; N)$, the quantities $\partial_m\tilde{\Phi}_\Lambda$ exist on $D^*$. It follows from (4.115) that

$$\partial_m\tilde{\delta}_s\Phi_\Lambda = \tilde{\delta}_s\partial_m\Phi_\Lambda = \partial_m\left[\left.\frac{\partial H_\Lambda}{\partial(\mathcal{O}_\Sigma{}^r\lambda^r\psi_r)}\right|_{\lambda=0}\left.\frac{\partial(\mathcal{O}_\Sigma{}^r\lambda^r\psi_r)}{\partial\lambda^r}\right|_{\lambda=0}\right] \tag{4.118}$$

are well defined on $D^*$.

We can now define function invariance. *A functional* $J[\{\Phi_\Lambda\}](L;D)$ *is said to be function invariant under the function substitution* $\Phi_\Lambda(x^m) \to \tilde{\Phi}_\Lambda(x^m; \eta^r)$ *if and only if*

$$\frac{dJ[\Phi_\Lambda](L;d)}{d\eta^s} = \lim_{\eta^s\to 0}\left[\frac{J[\{\Phi_\Lambda + \eta^s\tilde{\delta}_s\Phi_\Lambda\}](L;d) - J[\{\Phi_\Lambda\}](L;d)}{\eta^s}\right] \tag{4.119}$$

*vanishes for* $s = 1, \ldots, r$, *for all* $\{\psi_r(x^m)\} \in \mathcal{K}(r)$ *and for all* $d \subset D$. It is evident that the variation involved in (4.119) is a specialized version of the variation process comparing the images of two elements of $\mathfrak{D}_1(D^*; N)$ that are sufficiently close together in the norm $\mathfrak{D}_1(D^*; N)$ under the functional map $J[\{\Phi_\Lambda\}](L;d)$: we can use the results established in Sec. 4.2. When (4.46) is used, the following theorem is immediate.

**Theorem 4.9.** *A functional* $J[\{\Phi_\Lambda\}](L;D)$ *is function invariant under* $\Phi_\Lambda(x^m) \to \tilde{\Phi}_\Lambda(x^m; \eta^r)$ *if and only if*

$$\begin{aligned}&\frac{\partial L}{\partial\Phi_\Lambda}\left.\frac{\partial H_\Lambda}{\partial(\mathcal{O}_\Sigma{}^r\lambda^r\psi_r)}\right|_{\lambda=0}\left.\frac{\partial(\mathcal{O}_\Sigma{}^r\lambda^r\psi_r)}{\partial\lambda^s}\right|_{\lambda=0}\\ &+\frac{\partial L}{\partial(\partial_m\Phi_\Lambda)}\partial_m\left[\left.\frac{\partial H_\Lambda}{\partial(\mathcal{O}_\Sigma{}^r\lambda^r\psi_r)}\right|_{\lambda=0}\left.\frac{\partial(\mathcal{O}_\Sigma{}^r\lambda^r\psi_r)}{\partial\lambda^s}\right|_{\lambda=0}\right] = 0\end{aligned} \tag{4.120}$$

*is an identity on* $\mathcal{K}(r)$ *for* $s = 1, 2, \ldots, r$.

As an explicit example, consider the case where

$$\Phi_1 = f_1, \ldots, \Phi_n = f_n, \Phi_{n+1} = \varphi, \quad N = n + 1;$$

$$r = 1, \quad \mathcal{O}_1^{\ 1}\psi_1 = \partial_1\psi, \ldots, \quad \mathcal{O}_n^{\ 1}\psi_1 = \partial_n\psi, \ \mathcal{O}_{n+1}^1\psi_1 = e^{\psi}\ ;$$

$$H_1 = \Phi_1 + \mathcal{O}_1^{\ 1}\psi_1, \ldots, \ H_n = \Phi_n + \mathcal{O}_n^{\ 1}\psi_1\,, \quad H_{n+1} = \Phi_{n+1}\mathcal{O}_{n+1}^1\psi_1\,.$$

We then have

$$\tilde{\delta}_1\Phi_1 = \partial_1\psi_1, \ldots, \ \tilde{\delta}_1\Phi_n = \partial_n\psi_1, \ \tilde{\delta}_1\Phi_{n+1} = \Phi_{n+1}\psi_1$$

$$\partial_m\tilde{\delta}_1\Phi_1 = \partial_m\partial_1\psi_1, \ldots, \ \partial_m\tilde{\delta}_1\Phi_n = \partial_m\partial_n\psi_1,$$

$$\partial_m\tilde{\delta}_1\Phi_{n+1} = \psi_1\partial_m\Phi_{n+1} + \Phi_{n+1}\partial_m\psi\,,$$

and equations (4.120) are equivalent to the one equation

$$0 = \frac{\partial L}{\partial\Phi_\Lambda}\tilde{\delta}_1\Phi_\Lambda + \frac{\partial L}{\partial(\partial_m\Phi_\Lambda)}\partial_m(\tilde{\delta}_1\Phi_\Lambda)\,.$$

We accordingly have that

$$0 = \sum_{t=1}^{n}\frac{\partial L}{\partial\Phi_t}\partial_t\psi_1 + \frac{\partial L}{\partial\Phi_{n+1}}\Phi_{n+1}\psi_1 + \sum_{t=1}^{n}\frac{\partial L}{\partial(\partial_m\Phi_t)}\partial_m\partial_t\psi_1$$

$$+ \frac{\partial L}{\partial(\partial_m\Phi_{n+1})}(\psi_1\partial_m\Phi_{n+1} + \Phi_{n+1}\partial_m\psi_1)$$

must be an identity for all $\psi_1 \in C^2(D^*)$. Since $\partial_m\partial_t\psi_1 = \partial_t\partial_m\psi_1$, the above identity can be satisfied if and only if

$$0 = \Phi_{n+1}\frac{\partial L}{\partial\Phi_{n+1}} + (\partial_m\Phi_{n+1})\frac{\partial L}{\partial(\partial_m\Phi_{n+1})}\,, \tag{4.121}$$

$$0 = \frac{\partial L}{\partial\Phi_t} + \Phi_{n+1}\frac{\partial L}{\partial(\partial_t\Phi_{n+1})}\,, \quad 1 \le t \le n\,, \tag{4.122}$$

$$0 = \frac{\partial L}{\partial(\partial_m\Phi_t)} + \frac{\partial L}{\partial(\partial_t\Phi_m)}\,, \quad 1 \le (m, t) \le n\,. \tag{4.123}$$

Now, $L$ is a function of the $n + N + nN = n^2 + 3n + 1$ argument $x^m, \Phi_\Lambda, \partial_m\Phi_\Lambda$. An easy calculation shows that the general solution of

(4.123) is given by

$$L = L(x^m, \Phi_k, f_{km}, \Phi_{n+1}, \partial_k \Phi_{n+1}) \tag{4.124}$$

where

$$f_{km} = \partial_k \Phi_m - \partial_m \Phi_k, \quad 1 \le (k,m) \le n \tag{4.125}$$

The quantities $f_{km}$ thus satisfy the identities

$$\partial_k f_{mt} + \partial_m f_{tk} + \partial_t f_{km} = 0, \quad f_{km} = -f_{mk}. \tag{4.126}$$

The general solution of (4.121) is given by

$$L = L(x^m, \Phi_k, \partial_m \Phi_k, U_k), \quad U_k = \frac{\partial_k \Phi_{n+1}}{\Phi_{n+1}}. \tag{4.127}$$

Hence, we must have

$$L = L(x^m, \Phi_k, f_{km}, U_k). \tag{4.128}$$

From (4.128) and (4.127), $\dfrac{\partial L}{\partial(\partial_t \Phi_{n+1})} = \dfrac{\partial L}{\partial U_t} \dfrac{1}{\Phi_{n+1}}$; and accordingly, (4.122) gives

$$0 = \frac{\partial L}{\partial \Phi_t} + \frac{\partial L}{\partial U_t}, \quad 1 \le t \le n. \tag{4.129}$$

The general solution of this equation is given by

$$L = L(x^m, \partial_m \Phi_k, \xi_k), \quad \xi_k = \Phi_k - U_k. \tag{4.130}$$

Combining the results, we have finally

$$L = L(x^m, \xi_k, f_{km}) \tag{4.131}$$

where

$$\begin{cases} f_{km} = \partial_k \Phi_m - \partial_m \Phi_k, & 1 \le (k,m) \le n \\ \xi_k = \Phi_k - U_k, & U_k = \dfrac{\partial_k \Phi_{n+1}}{\Phi_{n+1}} \end{cases} \tag{4.132}$$

and

$$\partial_k f_{mt} + \partial_m f_{tk} + \partial_t f_{km} = 0\,, \quad f_{km} = -f_{mk}\,. \tag{4.133}$$

Let us compute the Euler equations with the lagrangian function determined in the above example. Set

$$F^{km} = \frac{\partial L}{\partial f_{km}}\,, \quad S^k = \frac{\partial L}{\partial \xi_k}\,, \tag{4.134}$$

then $F^{km} = -F^{mk}$ and

$$\frac{\partial L}{\partial(\partial_t \Phi_u)} = F^{km}\frac{\partial f_{km}}{\partial(\partial_t \Phi_u)} = F^{tu} - F^{ut} = 2F^{tu}\,, \quad \frac{\partial L}{\partial \Phi_u} = S^k \frac{\partial \xi_k}{\partial \Phi_u} = S^u\,,$$

$$\frac{\partial L}{\partial \Phi_{n+1}} = S^k \frac{\partial \xi_k}{\partial \Phi_{n+1}} = -S^k \frac{\partial U_k}{\partial \Phi_{n+1}} = +\frac{S^k}{(\Phi_{n+1})^2}\partial_k \Phi_{n+1}$$

$$\frac{\partial L}{\partial(\partial_t \Phi_{n+1})} = S^k \frac{\partial \xi_k}{\partial(\partial_t \Phi_{n+1})} = -S^k \frac{\partial U_k}{\partial(\partial_t \Phi_{n+1})} = -\frac{S^t}{\Phi_{n+1}}\,.$$

We thus have

$$0 = \{\mathcal{E}|L\}_{\Phi_u} = S^u - 2\partial_t F^{tu}\,, \quad 1 \le u \le n \tag{4.135}$$

$$0 = \{\mathcal{E}|L\}_{\Phi_{n+1}} = \frac{S^k \partial_k \Phi_{n+1}}{(\Phi_{n+1})^2} + \partial_t \frac{S^t}{\Phi_{n+1}} = \frac{1}{\Phi_{n+1}}\partial_t S^t\,. \tag{4.136}$$

Since $F^{km} = -F^{mk}$, the system (4.135) and (4.133) are the generalized Maxwell's equations in an $n$-dimensional manifold with the constitutive relations given by the first of (4.134). *Hence, Maxwell-like equations always result from the function invariance of* $J[\{\Phi_\Lambda\}](L;D)$ *under the function substitution* $f_k \to f_k + \partial_k \psi$. Further, (4.135) and (4.136) show that $S^t$ is the conserved current vector. *The function invariance considered above results in the full content of Maxwell's equations and automatically controls the fields in addition to* $f_k$ *in such a way that their Euler equations lead to a conserved current vector.* An important consequence of function invariance must now be obtained.

**Theorem 4.10.** *If* $J[\{\Phi_\Lambda\}](L;d)$ *is function invariant under*

$$\Phi_\Lambda \to \widetilde{\Phi}_\Lambda = H_\Lambda(\Phi_\Sigma, \mathcal{O}_\Sigma{}^r\lambda^r\psi_r) = \Phi_\Lambda + \eta^r\widetilde{\delta}_r\Phi_\Lambda + o(\|\eta\|) \qquad (4.137)$$

*and if* $\{\Phi_\Lambda\} \in S(L)$ *then* $\Phi_\Lambda + \eta^r\widetilde{\delta}_r\Phi_\Lambda \in S(L)$ *for all* $\|\eta\|$ *sufficiently small.*

**Proof.** Combining (4.137) with a function variation that vanishes on $\partial D$,

$$\Phi_\Lambda \to \Phi_\Lambda + \xi^\Gamma\delta_\Gamma\Phi_\Lambda + o(\xi^\Gamma),$$

we have

$$\Phi_\Lambda \to \widetilde{\widetilde{\Phi}}_\Lambda = \Phi_\Lambda + \xi^\Gamma\delta_\Gamma\Phi_\Lambda + \eta^r\widetilde{\delta}_r\Phi_\Lambda + o(\eta,\xi).$$

Hence the variation induced in $J[\{\Phi_\Lambda\}](L;D)$ is given by

$$\Delta J[\{\Phi_\Lambda\}](L;D) = \int_D \left\{ \frac{\partial L}{\partial\Phi_\Lambda}(\xi^\Gamma\delta_\Gamma\Phi_\Lambda + \eta^r\widetilde{\delta}_r\Phi_\Lambda) + \frac{\partial L}{\partial(\partial_m\Phi_\Lambda)}(\xi^\Gamma\partial_m\delta\ \Phi_\Lambda + \eta^r\partial_m\widetilde{\delta}_r\Phi_\Lambda)\right\} dV(x).$$

Now $\{\Phi_\Lambda\} \in S(L)$ implies that

$$\int_{D^*} \xi^\Gamma\left[\frac{\partial L}{\partial\Phi_\Lambda}\delta_\Gamma\Phi_\Lambda + \frac{\partial L}{\partial(\partial_m\Phi_\Lambda)}\partial_m\delta_\Gamma\Phi_\Lambda\right] dV(x).$$

vanishes, so that

$$\Delta J[\{\Phi_\Lambda\}](L;D) = \int \eta^r\left\{\frac{\partial L}{\partial\Phi_\Lambda}\widetilde{\delta}_r\Phi_\Lambda + \frac{\partial L}{\partial(\partial_m\Phi_\Lambda)}\partial_m\widetilde{\delta}_r\Phi_\Lambda\right\} dV(x).$$

However, the integral on the right-hand side vanishes by the hypothesis that $J[\{\Phi_\Lambda\}](L;D)$ is function invariant under (4.137) and hence $\Delta J[\{\Phi_\Lambda\}](L;D) = o(\xi,\eta)$.

Now,

$$\Delta J[\{\widetilde{\Phi}_\Lambda\}](L;D) = \Delta J[\{\Phi_\Lambda\}](L;D) + o(\eta)$$

for each element of $\mathcal{K}(r)$ and each $\{\eta^r\}$ such that $\|\eta\| \ll 1$. This follows from the fact that

$$\Phi_\Lambda \to \approx{\Phi}_\Lambda = \Phi_\Lambda + \xi^\Gamma \delta_\Gamma \Phi_\Lambda + \eta^r \tilde{\delta}_r \Phi_\Lambda + o(\eta, \xi)$$

is equivalent to

$$\Phi_\Lambda \to \Phi_\Lambda + \eta^r \tilde{\delta}_r \Phi_\Lambda + o(\eta) = \tilde{\Phi}_\Lambda + o(\eta) \to \tilde{\Phi}_\Lambda + \xi^\Gamma \delta_\Gamma \Phi_\Lambda + o(\xi, \eta)$$

and to the function invariance of $J[\{\Phi_\Lambda\}](L;D)$ under $\Phi_\Lambda \to \tilde{\Phi}_\Lambda$. We thus have

$$\Delta J[\{\tilde{\Phi}\}](L;D) = o(\xi)$$

for each $\{\psi_r\} \in \mathcal{K}(r)$ and each $\{\eta^r\}$ such that $\|\eta\| \ll 1$, and hence $\{\tilde{\Phi}_\Lambda\} \in S(L)$.

*Remark.* It can be shown that the above result holds for all finite norms of $\{\eta^k\}$ if the functions $H_\Lambda(\,;\,)$, defining the function substitution, form an additive group. This result is not established here for it is beyond the level of our analysis.

Theorem 4.10 states that an $r$-parameter function invariance embeds each solution of the Euler equations in an $r$-parameter family of solutions of the Euler equations. Thus, the solutions of the Euler equations are determined only to within the $r$-parameter family obtained under the function invariance, regardless of what boundary or initial conditions are imposed on the problem. In the case of Maxwell's equations, $f_i$ is determined only to within the one-parameter family $f_i + \partial_i \psi$, and hence $f_i$ is unique only after specification of at least one subsidiary condition. This causes many problems and, in the case of Maxwell's equations, reveals that the so-called gauge condition (Lorentz' gauge condition, etc.) is an independent assumption at the disposal of the investigator. Stated another way, if we have a lagrangian function that admits an $r$-parameter family of function invariances, then we must specify at least $r$ additional conditions (in addition to boundary and initial conditions) if we are to obtain a unique solution to the Euler equations.

## 4.8. LAGRANGIAN CLASSES

Strong invariance and invariance groups were studied in Sec. 4.5. There we assumed that the lagrangian function was given and obtained equations satisfied by the infinitesimal generators of a group in order that the group be an invariance group. The results obtained were not useful, however, for they demanded that certain expressions be identities on $\mathfrak{D}_1(D^*; N)$—the conditions could not be given explicitly for the general case. Reversing the problem, we assume that a given finite continuous group of transformations is given and seek all lagrangian functions which admit this group as either a coordinate invariance group or a point invariance group. Accordingly, we give the following definitions.

*If an $r$-parameter continuous group $G_r$ of coordinate transformations (point transformations) leaves $J[\{\Phi_\Lambda\}](L; d)$ invariant for all $d \subset D$ when $G_r$ is restricted to a sufficiently small neighborhood of the identity element, then $L$ is said to admit $G_r$ as a local coordinate (point) invariance group.*

*The collection of all lagrangian functions admitting $G_r$ as a local coordinate invariance group is referred to as the lagrangian coordinate class $L_c(G_r)$.*

*The collection of all lagrangian functions admitting $G_r$ as a local point invariance group is referred to as the lagrangian point class $L_p(G_r)$.*

The following theorem is easily established from (4.45) and (4.47) and the results of the previous sections.

**Theorem 4.11.** *A necessary and sufficient condition that $L \in L_c(G_r)$, in which $G_r$ is a given group with infinitesimal generators $v_a{}^m(x^k)$, is that $L$ satisfy the system of $r$ partial differential equations*

$$\begin{aligned} L\partial_m v_a{}^m + v_a{}^m \partial^*_m L + \underset{v_a}{D}\Phi_\Lambda \frac{\partial L}{\partial \Phi_\Lambda} \\ + (\partial_m \underset{v_a}{D}\Phi_\Lambda - \partial_m v_a{}^k \partial_k \Phi_\Lambda)\frac{\partial L}{\partial(\partial_m \Phi_\Lambda)} = 0, \quad a = 1, \dots, r. \end{aligned} \tag{4.138}$$

*A necessary and sufficient condition that $L \in L_p(G_r)$, in which $G_r$ is a given group with infinitesimal generators $v_a{}^m(x^k)$, is that $L$ satisfy the system of $r$ partial differential equations*

$$L\partial_m v_a{}^m + v_a{}^m \partial^*_m L - \partial_m v_a{}^k \partial_k \Phi_\Lambda \frac{\partial L}{\partial(\partial_m \Phi_\Lambda)} = 0, \quad a = 1, \dots, r. \tag{4.139}$$

We have finally arrived at the formulation of invariance requirements (coordinate and point) which is compatible with physical information—we assume knowledge of a group $G_r$ which results in invariance structure and then seek the possible functional dependences of the lagrangian function for which the functionals $J[\{\Phi_\Lambda\}](L;\cdot)$ will be invariant. We must also know the structure of the $N$-tuple of functions that constitute elements of $\mathfrak{D}_1(D^*;N)$—we must know the geometry class to which $\{\Phi_\Lambda(x^m)\}$ belongs—since (4.138) involves the quantities $\underset{v_a}{D}\Phi_\Lambda$ and these quantities are only determined once the transformation properties of $\Phi_\Lambda(x^m)$ are known. We must have *a priori* knowledge that the lagrangian function depends on scalars, vectors, tensors or combinations of these, together with their derivatives. This is obvious, for the possible equations which a scalar field can satisfy are intrinsically different from those which a vector field can satisfy. These different possibilities will lead to different invariance structures.

We can now establish the relation between lagrangian classes and invariance transformations.

**Theorem 4.12.** *An r-parameter group $G_r$ is a coordinate invariance group of $J[\{\Phi_\Lambda\}](L;D)$ if and only if $L \in L_c(G_r)$. An r-parameter group $G_r$ is a point invariance group of $J[\{\Phi_\Lambda\}](L;D)$ if and only if $L \in L_p(G_r)$.*

**Proof.** We shall give a detailed proof of the first statement. The proof of the second statement is identical with appropriate changes made for the interchange of coordinate transformations and point transformations. By Theorem 4.12, $L \in L_c(G_r)$ if and only if $L$ is a solution of the system of partial differential equations

$$L\,\partial_m v_a{}^m + v_a{}^m \partial_m^* L + \underset{v_a}{D}\Phi_\Lambda \frac{\partial L}{\partial \Phi_\Lambda} + (\partial_m \underset{v_a}{D}\Phi_\Lambda - \partial_m v_a{}^k \partial_k \Phi_\Lambda)\frac{\partial L}{\partial(\partial_m \Phi_\Lambda)} = 0\,. \tag{4.140}$$

By equation (4.68) of Theorem 4.4 on invariance groups, $G_r$ is an invariance group of $J[\{\Phi_\Lambda\}](L;D)$ if and only if

$$W_k{}^m \partial_m v_a{}^k = v_a{}^k \partial_k^* L + \underset{v_a}{D}\Phi_\Lambda \frac{\partial L}{\partial \Phi_\Lambda} + \partial_m \underset{v_a}{D}\Phi_\Lambda \frac{\partial L}{\partial(\partial_m \Phi_\Lambda)} \tag{4.141}$$

is identically satisfied over $\mathfrak{D}_1(D^*;N)$. Supplose that $L$ is a solution of (4.140) and set

$$\Gamma_a = -W_k{}^m \partial_m v_a{}^k + v_a{}^k \partial_k^* L + \underset{v_a}{D}\Phi_\Lambda \frac{\partial L}{\partial \Phi_\Lambda} + \partial_m \underset{v_a}{D}\Phi_\Lambda \frac{\partial L}{\partial(\partial_m \Phi_\Lambda)}\,. \tag{4.142}$$

When (4.140) is solved for $v_a{}^m \partial_k^* L$ and the result is substituted into (4.142), we have

$$\Gamma_a = \left[ -W_k{}^m + \partial_k \Phi_\Lambda \frac{\partial L}{\partial(\partial_m \Phi_\Lambda)} - L \delta_k{}^m \right] \partial_m v^k . \tag{4.143}$$

For any $\{\Phi_\Lambda\} \in \mathfrak{D}_1(D; N)$, (4.37) shows that the expression within the parentheses on the right-hand side of (4.143) vanishes, and hence, $\Gamma_a = 0$ holds as an identity over $\mathfrak{D}_1(D^*; N)$. A comparison of (4.142) and (4.141) then shows that (4.141) is an identity over $\mathfrak{D}_1(D; N)$. Conversely, suppose that $v_a{}^k(x^m)$ is such that (4.141) is identically satisfied over $\mathfrak{D}_1(D^*; N)$ and set

$$\begin{aligned} \Sigma_a &= L \partial_m v_a{}^m + v_a{}^m \partial_m^* L + \underset{v_a}{D} \Phi_\Lambda \frac{\partial L}{\partial \Phi_\Lambda} \\ &+ (\partial_m \underset{v_a}{D} \Phi_\Lambda - \partial_m v_a{}^k \partial_k \Phi_\Lambda) \frac{\partial L}{\partial(\partial_m \Phi_\Lambda)} . \end{aligned} \tag{4.144}$$

When (4.141) is solved for $v_a{}^k \partial_k^* L$ and the result is substituted into (4.144), we obtain

$$\Sigma_a = \left[ W_k{}^m - \partial_k \Phi_\Lambda \frac{\partial L}{\partial(\partial_m \Phi_\Lambda)} + L \delta_k{}^m \right] \partial_m v_a{}^k \tag{4.145}$$

for all $\{\Phi_\Lambda(x^m)\} \in \mathfrak{D}_1(D^*; N)$. In view of (4.37), the quantity within the parentheses vanishes for all $\{\Phi_\Lambda(x^m)\} \in \mathfrak{D}_1(D^*; N)$, and hence $\Sigma_a = 0$ is an identity over $\mathfrak{D}_1(D^*; N)$. It then follows from (4.144) and (4.140), that (4.140) holds identically over $\mathfrak{D}_1(D^*; N)$, while $v_a{}^k(x^m)$ are given functions which constitute a complete solution of (4.141) when this equation holds identically over $\mathfrak{D}_1(D^*; N)$. It thus follows that $L$ must satisfy the system of $r$ equations (4.140) as a function of the $n + N + nN$ arguments $x^m, \Phi_\Lambda, \partial_m \Phi_\Lambda$, and hence $L \in L_c(G_r)$.

We now have the necessary results by which we can handle the physics of observables. The results of Chaps. 1 and 2 show that we can embed any system of equations in a variational statement with no loss of generality, although the embedding may involve subsidiary (unphysical) functions. Thus, if we know the governing equations of a physical system, we can embed them and use the results of Chap. 4 to obtain their invariance structure and the resulting conservation laws of the system—the generalized symmetries of the physical system.

In actual fact, however, the governing equations of a physical system are never known with certainty—only systems of equations which yield predictions consistent with observations to within the accuracy of measurability of the observables are known. The primary source of this lack of knowledge stems from the fact that the governing equations of a physical system involve constitutive relations describing the inherent relations between the intrinsic and extrinsic quantities of physical system and the extrinsic quantities are basically immeasurable. However, suppose that, regardless of the form of the governing equations, we know a group that leaves the physics invariant in content—we know that the physics admits a group of symmetries. By the results of Sec. 4.5 (equations (4.72) and (4.73) and Theorem 4.5), this is equivalent to knowing a system of conservation laws of the physical system. We can then use the results established here, provided that we know that the primary physical quantities belong to a definite geometry class—the physics involves a definite collection of scalars, vectors, tensors, spinors, etc.—and obtain the most general lagrangian function for which the given system of symmetries is exhibited. We thus determine the most general system of Euler equations obtainable under the condition that the physical system exhibit the given symmetries. This gives the set of all possible governing equations for physical systems that can be described by elements of the assumed geometry class and that admits the known symmetries. Once we have this, we can inspect the set and see if we can obtain a more definite system of governing equations by the addition of new conditions. If we add the condition that the sum of any two solutions is always a solution, then the equations must be linear; and hence the lagrangian can at most be quadratic. In this instance we obtain a fairly simple system of equations which can be viewed as the linearized approximation of the complete system obtained. The important thing to note here is that we do not establish preconceived prejudices. We obtain the most general lagrangian (Euler equations) compatible with the physical symmetries and, thus, select physical systems from a wide area. Argument can then be made from the general to the particular with exactness and economy. The result is a practical and theoretically sound method for examining physical phenomena by means of mathematical models.

Let us look at the example considered in Sec. 4.5 from this point of view. We have $N = 1$, $n = 2$, $\Phi_\Lambda$ a scalar function,

$$G_3: \begin{cases} v_1{}^x = 1, & v_2{}^x = 0, \quad v_3{}^x = t \\ v_1{}^t = 0, & v_2{}^t = 1, \quad v_3{}^t = a^2 x \end{cases}$$

and $\underset{v_a}{D}\Phi_1 = 0$ since $\Phi_1$ is a scalar. This last condition results in $L_c(G_3) \equiv L_p(G_3)$ since (4.138) and (4.139) are then equivalent. We accordingly have, by (4.139), that $L \in L_c(G_3)$ or $L \in L_p(G_3)$ if and only if

$$a = 1: \quad \frac{\partial^* L}{\partial x} = 0 \tag{4.146}$$

$$a = 2: \quad \frac{\partial^* L}{\partial t} = 0 \tag{4.147}$$

$$a = 3: \quad t\frac{\partial^* L}{\partial x} + a^2 x\frac{\partial^* L}{\partial t} - \partial_x \Phi_1 \frac{\partial L}{\partial(\partial_t \Phi_1)} - a^2 \partial_t \Phi_1 \frac{\partial L}{\partial(\partial_x \Phi_1)} = 0 \tag{4.148}$$

The conditions (4.146) and (4.147) yield

$$L = \overline{L}(\Phi_1, \partial_x \Phi_1, \partial_t \Phi_1) . \tag{4.149}$$

Substituting this result into (4.148) gives

$$\partial_x \Phi_1 \frac{\partial \overline{L}}{\partial(\partial_t \Phi_1)} + a^2 \partial_t \Phi_1 \frac{\partial \overline{L}}{\partial(\partial_x \Phi_1)} = 0 . \tag{4.150}$$

The general solution of this first-order linear partial differential equation is given by

$$\overline{L} = F[\Phi_1, (\partial_x \Phi_1)^2 - a^2(\partial_t \Phi_1)^2] .$$

Hence $L \in L_c(G_3)$, $L \in L_p(G_3)$ if and only if

$$L = F[\Phi_1, (\partial_x \Phi_1)^2 - a^2(\partial_t \Phi_1)^2] , \tag{4.151}$$

in which $F$ is any function of two variables of class $C^2$. We thus began with a lagrangian function with five arguments $x, t, \Phi_1, \partial_x \Phi_1, \partial_t \Phi_1$ and ended with a lagrangian function specified in terms of a function of only two known arguments, $\Phi_1$ and $(\partial_x \Phi_1)^2 - a^2(\partial_t \Phi_1)^2$. Under the requirement $L \in L_c(G_3)$ or $L \in L_p(G_3)$, a reduction of three in the dimension of the arguments of the lagrangian function obtains. If we require that the resulting Euler equations be linear, then we must have

$$L = \bar{a}(\Phi_1)^2 + b\Phi_1 + c + d[(\partial_x \Phi)^2 - a^2(\partial_t \Phi)^2]$$

in which $\bar{a}$, $b$, $c$ and $d$ are constants and in which the $c$ can be omitted since $c \in \mathfrak{N}(1)$. The resulting Euler equation is

$$0 = \{\mathcal{E}|L\}_{\Phi_1} = 2\bar{a}\Phi_1 + b - 2d(\partial_x\partial_x\Phi_1 - a^2\partial_t\partial_t\Phi_1) .$$

We have thus established the following conclusion: *if* (1) $L \in L_c(G^3)$ *or* $L \in L_p(G^3)$, (2) *the Euler equations are linear and homogeneous*, (3) $N = 1$, $n = 2$, *and* $\Phi_1$ *is a scalar function, then* $\Phi_1$ *satisfies the Klein-Gordon equation.*

We can now give a partial answer to the problems caused by the null class of the Euler-Lagrange operator being nonvacuous, as they pertain to invariance or lack of invariance. We assume that $\{\Phi_\Lambda\}$ belongs to a given geometry class $\mathcal{G}(p,N;F_\Lambda)/G_r$ for all $\{\Phi_\Lambda\} \in \mathfrak{D}_1(D^*;N)$ and define the class $\mathfrak{N}(N)/G_r$, *the restriction of* $\mathfrak{N}(N)$ *to* $G_r$, *by*

$$\frac{\mathfrak{N}(N)}{G_r} = \left[\mathfrak{N}(N) \cap \frac{\mathcal{G}(p,N,F_\Lambda)}{G_r}\right] \cap L_c(G_r) . \tag{4.152}$$

We then have $(L + A) \in L_c(G_r)$ for any $A \in \mathfrak{N}(N)/G_r$, since the equations defining $L_c(G_r)$, namely (4.138), are linear in $L$. It thus follows that $(L + A)$ admits the same coordinate invariance group as $L$ for all $A \in \mathfrak{N}(N)/G_r$. It is also evident, from the linearity of (4.138) in $L$, that if $A \in \mathfrak{N}(N)$ but is not contained in $\mathfrak{N}(N)/G_r$, then $L + A$ will not admit the same coordinate invariance group as $L$. We thus have the following result.

**Theorem 4.13.** *If* $L$ *admits a coordinate invariance group* $G_r$, *then* $L + A$ *admits the same coordinate invariance group if and only if* $A \in \mathfrak{N}(N)/G_r$.

**Corollary.** *Let* $L$ *admit a coordinate invariance group* $G_r$ *and yield a given system of Euler equations. If* $\mathfrak{N}(N)/G_r$ *does not exhaust* $\mathfrak{N}(N) \cap \mathcal{G}(p,N;F_\Lambda)/G_r$, *in which the argument of* $L$ *belong to* $\mathcal{G}(p,N;F_\Lambda)/G_r$, *then the class of all lagrangian functions* $\hat{\mathfrak{L}}$ *which yields the given system of Euler equations cannot all admit* $G_r$ *as a coordinate invariance group. In fact, the coordinate invariance group admitted by all elements of* $\hat{\mathfrak{L}}$ *can be no more than a subgroup of* $G_r$ *and consists of the coordinate invariance group which generates the transformations common to the intersection of the lagrangian coordinate classes of each of the elements of* $\hat{\mathfrak{L}}$.

The last statement follows from the fact that each lagrangian coordinate class spans all solutions of the system of equations (4.138).

Results of an identical nature can be obtained for point invariance groups. Their formulation and proof are simple and are left to the student.

## 4.9. GEOMETRY CLASSES AND COORDINATE INVARIANCE OF $S(L)$

If we start with a given lagrangian function, we can obtain the elements of $S(L)$ that stationarize $J[\{\Phi_\Lambda\}](L;D)$ by solving the Euler equations: $S(L)$ is the set of all elements of $\mathfrak{D}_1(D^*;N)$ for which

$$J[\{\Phi_\Lambda\}](L;D) = \int_D L[x^m, {}_\alpha\Phi_\Lambda(x^m)]\,dV(x) \tag{4.153}$$

is stationary. Suppose that we subject $D^*$ to the action of a continuous $r$-parameter group $G_r$ of coordinate transformations, and then consider the functional

$$\tilde{J}[\{\Phi_\Lambda\}](L;D) = \int_D L[x^{m'}, {}_\alpha\Phi_{\Lambda'}(x^m)]\,dV(x') \,. \tag{4.154}$$

If we express $x'$ and ${}_\alpha\tilde{\Phi}_{\Lambda'}$ in terms of $x$ and ${}_\alpha\Phi_\Lambda$ by the transformation equation of $G_r$ and the laws of transformation for ${}_\alpha\Phi_\Lambda$, where the $\{\Phi_\Lambda\}$ are assumed to be elements of a geometry class $\mathcal{G}(p,N;F_\Lambda)/G_r$, we have

$$\begin{aligned}\tilde{J}[\{\Phi_\Lambda\}](L;D) &= \int_D L[x^{m'}, {}_\alpha\Phi_{\Lambda'}(x^m)]\left(\frac{x'}{x}\right)dV(x)\\ &= \int_D L'(x^m, {}_\alpha\Phi_\Lambda)\,dV(x) = J[\{\Phi_\Lambda\}](L';D)\end{aligned} \tag{4.155}$$

where

$$L'(x^m, {}_\alpha\Phi_\Lambda) = L[x^{m'}, {}_\alpha\Phi_{\Lambda'}(x^m)]\left(\frac{x'}{x}\right) . \tag{4.156}$$

A coordinate transformation is thus equivalent to a new lagrangian function $L'$. Since the elements of the stationarization set depend in an intrinsic fashion on which lagrangian function is used, the question arises as to whether $S(L)$ and $S(L')$ are identical. This qustion is equivalent to the question of whether the image of an element of $S(L)$ under $G_r$ will render $J[\{\Phi_\Lambda\}](L;D)$ stationary in value. Stated another way, *will every image of* $S(L)$ *under* $G_r$ *be such that it will satisfy the image of the Euler equations under the transformation induced by* $G_r$?

We already know that $J[\cdot](L;d)$ and $\tilde{J}[\cdot](L;d) = J[\cdot](L';d)$ agree for $G_r$ restricted to a sufficiently small neighborhood of the identiy and for all $d \subset D$ if and only if $L \in L_c(G_r)$. This follows directly from the considerations of the previous sections, in particular, from Secs. 4.2 and 4.8. We accordingly must study the class $L_c(G_r)$.

As previously noted the arguments of the lagrangian function are the $N$ quantities $\Phi_\Lambda$, the $nN$ quantities $\partial_m \Phi_\Lambda$, and the $n$ quantities $x^m$. In order to simplify the resulting analysis, let us define the quantities $\psi^A$, $A = 1, \ldots, n(N+1) + N$, by

$$\psi^\Lambda = \Phi_\Lambda, \quad \psi^{n+\Lambda} = \partial_1 \Phi_\Lambda, \ldots, \psi^{nN+\Lambda} = \partial_n \Phi_\Lambda, \quad \psi^{N(n+1)+m} = x^m, \tag{4.157}$$

so that we may write

$$L = L(\psi^A). \tag{4.158}$$

Thus, $L$ may be considered as a function defined over $[n(N+1)+N]$-dimensional space $X$ with Cartesian coordinates $\psi^A$. Further, let us define the functions $J_a{}^A$, $a = 1, \ldots, r$, by

$$\left\{\begin{aligned} J_a{}^A &= \underset{v_a}{\Delta} \Phi_\Lambda, \\ J_a{}^{N+\Lambda} &= \partial_1(\underset{v_a}{\Delta} \Phi_\Lambda) - \partial_k \Phi_\Lambda \partial_1 v_a{}^k, \\ J_a{}^{2N+\Lambda} &= \partial_2(\underset{v_a}{\Delta} \Phi_\Lambda) - \partial_k \Phi_\Lambda \partial_2 v_a{}^k, \ldots, \\ J_a{}^{nN+\Lambda} &= \partial_n(\underset{v_a}{\Delta} \Phi_\Lambda) - \partial_k \Phi_\Lambda \partial_n v_a{}^k, \\ J_a{}^{N(n+1)+m} &= v_a{}^m, \end{aligned}\right. \tag{4.159}$$

in which $v_a{}^m(x^k)$ are the infinitesimal generators of $G_r$. The following lemma will be needed in what follows.

**Lemma 4.1.** *The functions $J_a{}^A$ depend only on the $n(N+1)+N$ quantities $\psi^A$, and hence are functions on the space $X$.*

**Proof.** Since $v_a{}^k(x^m)$ are given functions, the only terms occurring in (4.159) for which the result is not obvious are $\underset{v_a}{\Delta}\Phi_\Lambda$ and $\partial_m \underset{v_a}{\Delta}\Phi_\Lambda$. From (4.43) we have

$$\underset{v_a}{\Delta}\Phi_\Lambda = \partial_m\Phi_\Lambda\cdot v_a{}^m - \underset{v_a}{\pounds}\Phi_\Lambda .$$

From the known form of the Lie-derivative of a geometric object, the term $\partial_m\Phi_\Lambda\cdot v_a{}^m$ exactly cancels the corresponding term in $\underset{v_a}{\pounds}\Phi_\Lambda$. Hence $\underset{v_a}{\Delta}\Phi_\Lambda$ is defined on $X$ and is independent of $\partial_m \underset{v_a}{\Delta}\Phi_\Lambda$ for $m=1,\ldots,n$. It then follows that $\partial_m \underset{v_a}{\Delta}\Phi_\Lambda$ is defined on $X$ since we know that Lie differentiation and partial differentiation commute.

**Corollary.** *The quantities*

$$X_a = J_a{}^A\partial_A \qquad \left(= J_a{}^A\frac{\partial}{\partial\psi^A}\right) \tag{4.160}$$

*are well defined linear differential operators whose domain consists of all $C^1$ functions defined on the space $X$ with coordinate $\psi^A$.*

**Theorem 4.14.** $L \in L_c(G_r)$ *if and only if* $L$ *is a solution of the system of* $r$ *linear first-order partial differential equations*

$$(\partial_m v_a{}^m + X_a)F(\psi^A) = 0 , \quad a = 1,\ldots,r , \tag{4.161}$$

*where $v_a{}^m$ are the infinitesimal generators of $G_r$.*

**Proof.** This is merely a restatement of (4.138) in terms of the new symbols introduced here.

Theorem 4.14 reduces the problem of finding lagrangian functions belong $L_c(G_r)$ to the problem of determining common integrals of a system of linear partial differential equations of the first order and places these known results at our disposal.

**Theorem 4.15.** *Let $G_r$ be an $r$-parameter group of coordinate transformation such that $\partial_m v_a{}^m = 0$ for all $a=1,\ldots,r$, let $R$ denote the rank of the matrix with entries $J_a{}^A$ and let $R+C$ denote the order of the complete system obtained from $X_aF=0$. Then, either $n(N+1)+N \le R+C$, in which case $F$ is a constant function on $X$, or $L_c(G_r)$ consists of all $C^2$ functions of the $n(N+1)+N-C-R$ primitive integrals of the system $X_aF=0$.*

We now proceed to dispense with the condition that $G_r$ be volume-preserving, i.e., that $\partial_m v_a{}^m = 0$. Consider the transformation

$$F(\psi^A) = \rho(x^m) f(\psi^A) . \tag{4.162}$$

If we substitute (4.162) into (4.161) and use (4.159) and (4.158) we have

$$f(\psi^A)(v_a{}^m \partial_m + \partial_m v_a{}^m) \rho(x^k) + \rho(x^k) X_a f(\psi^A) = 0 . \tag{4.163}$$

**Theorem 4.16.** *Let the hypotheses of Theorem 4.15 be satisfied and let $\rho$ be a function of $x^m$ such that*

$$(v_a{}^m \partial_m + \partial_m v_a{}^m) \rho(x^k) = 0 \tag{4.164}$$

*holds. Then, either $n(N+1) + N \leq C + R$, in which case $L_c(G_r)$ consists only of constant multiples of $\rho$, or $L_c(G_r)$ consists of all products of $\rho$ and all $C^2$ functions of the $n(N+1) + N - R - C$ primitive integrals of the system $X_a f = 0$.*

In either event, there is a significant dimension reduction, following from the fact that the primitive integrals of a given system of first-order equations can be obtained explicitly. Therefore, in order to determine lagrangian functions belonging to $L_c(G_r)$, we require knowledge of a function of only $n(N+1) + N - R - C$ variables rather than a function of $n(N+1) + N$ variables.

**Theorem 4.17.** *If $L \in L_c(G_r)$, then*

$$\underset{v_a}{\pounds} L = \partial_m (L v_a{}^m) \tag{4.165}$$

*is an identity on the space $X$ with coordinates $\psi^A$.*

**Proof.** If $L \in L_c(G_r)$, then $L$ satisfies (4.161). When the explicit evaluation of $\underset{v_a}{\Delta} \Phi_\Lambda$ is used, together with (4.157) and (4.159), then (4.161) becomes the identity

$$\underset{v_a}{\pounds} (\Phi_\Lambda) \frac{\partial L}{\partial \Phi_\Lambda} + \underset{v_a}{\pounds} (\partial_m \Phi_\Lambda) \frac{\partial L}{\partial(\partial_m \Phi_\Lambda)} = \partial_m (L v_a{}^m) \tag{4.166}$$

on $X$. Now, $\underset{v_a}{\pounds} L = (\underset{v_a}{\pounds} x^m) \partial_m^* L + \underset{v_a}{\pounds} (\Phi_\Lambda) \frac{\partial L}{\partial(\Phi_\Lambda)} + \underset{v_a}{\pounds} (\partial_m \Phi_\Lambda) \frac{\partial L}{\partial(\partial_m \Phi_\Lambda)}$, since $L$ is defined on $X$, while $\underset{v_a}{\pounds} x^m = 0$ for the geometric objects $x^m$. The identity (4.166) thus implies the identity (4.165).

The reader should note that if $u^m(x^k)$ are the components of a contravariant vector that cannot be written as a linear combination of the $r$ contravariant vectors $\{v_a{}^m(x^k)\}$ with constant coefficients, then $\underset{u}{\pounds} L$ is in general different from $\partial_m(Lu^m)$. Referring to the example in Sec. 4.8, the lagrangian given by (4.166) is such that $\underset{v_a}{\pounds} L = \partial_m(Lv_a{}^m)$ for the three generators of the $G_3$, considered there, but $\underset{u}{\pounds} L \neq \partial_m(Lu^m)$ for $u^x = x^2 + 2t - 1$, $u^t = t^2 - \sqrt{3x} - te^{x^2}$.

**Theorem 4.18.** *If* $L \in L_c(G_r)$, *then* $L$ *transforms as a density or a* $\Delta$*-density of weight* $+1$ *under all elements of* $G_r$, *not just those in a sufficiently small neighborhood of the identity element.*

**Proof.** By Theorem 4.17, we have

$$\underset{v_a}{\pounds} L = \partial_m(Lv_a{}^m)$$

as an identity on $X$. Now, the Lie derivative of a quantity $L$ with respect to a vector $\{v_a{}^m(x^k)\}$ has the evaluation $\partial_m(Lv_a{}^m)$ if and only if $L$ transforms as a scalar density or a $\Delta$-density of weight $+1$ under the transformations $x^{m'} = x^m + tv_a{}^m(x^k) + o(t)$. Accordingly, $L$ transforms as a scalar density under each of the infinitesimal transformations of $G_r$. Since a scalar density of a $\Delta$-density of weight $+1$ is a linear geometric object, we know from the results of Chap. 3 that

$$(\underset{v_a}{\pounds}, \underset{v_b}{\pounds})L = C_{ab}{}^e \underset{v_c}{\pounds} L$$

holds identically in $X$. From this and the properties of continuous $r$-parameter groups, it follows that $L$ transforms as a scalar density or a $\Delta$-density of weight $+1$ under all elements of $G_r$.

Returning again to the example of Sec. 4.8, the lagrangian function (4.151),

$$L = L[\Phi_1, (\partial_x \Phi_1)^2 - a^2(\partial_t \Phi_1)^2]$$

transforms as a scalar density under the group $G_3$ considered in Sec. 4.8, but obviously does not transform as a scalar density under the one-parameter group $G_1$ generated by $u^x = t$, $u^t = -x$..

**Corollary.** *If* $L \in L_c(G_r)$, *we have*

$$L'(x^m, {}_{\boldsymbol{a}}\Phi_\Lambda) = L[x^{m'}, {}_{\boldsymbol{a}}\Phi_{\Lambda'}(x^m)]\left(\frac{x'}{x}\right) = L(x^m, {}_{\boldsymbol{a}}\Phi_\Lambda) \tag{4.167}$$

*for all elements of* $G_r$.

**Proof.** By Theorem 4.18, $L \in L_c(G_r)$ imples that $L$ transforms as a density or a $\Delta$-density of weight $+1$ under all elements of $G_r$. This means that

$$L[x^{m'}, {}_\alpha\Phi_{\Lambda'}(x^m)] = \left(\frac{x}{x'}\right) L(x^m, {}_\alpha\Phi_\Lambda) \tag{4.168}$$

since $|(x/x')| = (x/x')$ because every element of $G_r$ is continuously connected to the identity element. Further $(x/x') \neq 0$, and hence (4.168) and (4.156) combine to give (4.167).

We now have all the results required to answer the question raised at the beginning of Sec. 4.9.

**Theorem 4.19.** *If* $L \in L_c(G_r)$ *and the arguments of* $L$ *belong to a geometry class* $\mathcal{G}(p, N; F_\Lambda)/G_r$, *then, for any element of* $G_r$, $S(L)$ *and* $S(L')$ *are isomorphic under the map induced by the transformation laws of the geometry class* $\mathcal{G}(p, N; F_\Lambda)/G_r$.

**Proof.** By the previous corollary, we have $L' = L$ as an identity in $X$, and hence $S(L) = S(L')$. It is illustrative to give a direct proof, since we will thereby obtain the transformation law induced by the transformation laws of the given geometry class $\mathcal{G}(p, N; F_\Lambda)/G_r$. By (4.167) we have

$$L(x^{m'}, {}_\alpha\Phi_{\Lambda'})\left(\frac{x'}{x}\right) = L(x^m, {}_\alpha\Phi_\Lambda) .$$

Operating on both sides of the identity with $\{\mathcal{E}|\cdot\}_{\Phi_\Lambda}$ gives

$$\left\{\mathcal{E}|L(x^{m'}, {}_\alpha\Phi_{\Lambda'})\left(\frac{x'}{x}\right)\right\}_{\Phi_\Lambda} = \{\mathcal{E}|L\}_{\Phi_\Lambda} . \tag{4.169}$$

Since $\{\Phi_\Lambda\} \in \mathcal{G}(p, N; F_\Lambda)/G_1$, we have

$$\Phi_{\Lambda'} = F_\Lambda(\Phi_\Lambda, x^m, x^{m'}) \tag{4.170}$$

and

$$\partial_{m'}\Phi_{\Lambda'} = \partial_k \Phi_\Sigma A_{m'}{}^k \frac{\partial F_\Lambda}{\partial \Phi_\Sigma} + A_{m'}{}^k \partial_k^* F_\Lambda + \partial_{m'}^* F_\Lambda \tag{4.171}$$

where $A_{m'}{}^k = \partial x^k/\partial x^{m'}$, $A_m{}^{k'} = \partial x^{k'}/\partial x^m$. From (4.170) we easily obtain that

$$\frac{\partial \Phi_{\Lambda'}}{\partial(\partial_m \Phi_\Sigma)} = 0 ,$$

and hence

$$\left\{\mathcal{E}|L(x^{m'},{}_\alpha\Phi_{\Lambda'})\left(\frac{x'}{x}\right)\right\}_{\Phi_\Lambda} = \left(\frac{x'}{x}\right)\frac{\partial L(x^{m'},)}{\partial(\Phi_{\Lambda'})}\frac{\partial\Phi_{\Sigma'}}{\partial\Phi_\Lambda} + \left(\frac{x'}{x}\right)\frac{\partial L(x^{m'},)}{\partial(\partial_{k'}\Phi_{\Lambda'})}\frac{\partial(\partial_{k'}\Phi_{\Sigma'})}{\partial(\Phi_\Lambda)}$$

$$- \partial_m\left\{\left(\frac{x'}{x}\right)\frac{\partial L(x^{m'},)}{\partial(\partial_{k'}\Phi_{\Lambda'})}\frac{\partial(\partial_{k'}\Phi_{\Sigma'})}{\partial(\partial_m\Phi_\Lambda)}\right\}$$

$$= \left(\frac{x'}{x}\right)\left\{\frac{\partial L(x^{m'},)}{\partial\Phi_{\Sigma'}}\frac{\partial\Phi_{\Sigma'}}{\partial\Phi_\Lambda} - \partial_m\left[\frac{\partial L(x^{m'},)}{\partial(\partial_{k'}\Phi_{\Sigma'})}\right]\frac{\partial(\partial_{k'}\Phi_{\Sigma'})}{\partial(\partial_m\Phi_\Lambda)}\right.$$

$$+ \frac{\partial L(x^{m'},)}{\partial(\partial_{k'}\Phi_{\Sigma'})}\left\{\frac{\partial(\partial_{k'}\Phi_{\Sigma'})}{\partial\Phi_\Lambda} - \partial_m\left[\frac{\partial(\partial_{k'}\Phi_{\Sigma'})}{\partial(\partial_m\Phi_\Lambda)}\right]\right.$$

$$\left.\left. - \frac{\partial(\partial_{k'}\Phi_{\Sigma'})}{\partial(\partial_m\Phi_\Lambda)}\partial_m \ln\left(\frac{x'}{x}\right)\right\}\right\} . \tag{4.172}$$

A straightforward calculation based on (4.170) and (4.171) shows that

$$\frac{\partial(\partial_{k'}\Phi_{\Sigma'})}{\partial(\partial_m\Phi_\Lambda)} = A_{k'}{}^m\frac{\partial\Phi_{\Sigma'}}{\partial\Phi_\Lambda}$$

and

$$\frac{\partial(\partial_{k'}\Phi_{\Sigma'})}{\partial(\Phi_\Lambda)} = \partial_m\left[\frac{\partial(\partial_{k'}\Phi_{\Sigma'})}{\partial(\partial_m\Phi_\Lambda)}\right] + \frac{\partial(\partial_{k'}\Phi_{\Sigma'})}{\partial(\partial_m\Phi_\Lambda)}\partial_m \ln\left(\frac{x'}{x}\right).$$

Substituting these results into (4.172), we obtain

$$\left\{\mathcal{E}|L(x^{m'},{}_\alpha\Phi_{\Lambda'})\left(\frac{x'}{x}\right)\right\}_{\Phi_\Lambda} = \left(\frac{x'}{x}\right)\frac{\partial\Phi_{\Sigma'}}{\partial\Phi_\Lambda}\left\{\frac{\partial L(x^{m'},)}{\partial\Phi_{\Sigma'}} - \partial_{k'}\left[\frac{\partial L(x^{m'},)}{\partial(\partial_{k'}\Phi_{\Sigma'})}\right]\right\} \tag{4.173}$$

$$= \left(\frac{x'}{x}\right)\frac{\partial\Phi_{\Sigma'}}{\partial\Phi_\Lambda}\{\mathcal{E}|L(x^{m'},{}_\alpha\Phi_{\Lambda'})\}_{\Phi_{\Sigma'}} .$$

A combination of (4.173) and (4.169) thus gives us

$$\left(\frac{x'}{x}\right)\frac{\partial\Phi_{\Sigma'}}{\partial\Phi_{\Lambda}}\{\mathcal{E}|L(x^{m'},{}_{\alpha}\widetilde{\Phi_{\Lambda'}})\}_{\Phi_{\Sigma'}} = \{\mathcal{E}|L(x^{m},{}_{\alpha}\Phi_{\Lambda})\}_{\Phi_{\Lambda}}. \tag{4.174}$$

However,

$$J[\{\Phi_{\Lambda}\}](L;D) = \int_D L(x^m,{}_{\alpha}\Phi_{\Lambda})\,dV(x) \Rightarrow S(L)$$

$$\breve{J}[\{\Phi_{\Lambda}\}](L;D) = \int_D L(x^{m'},{}_{\alpha}\widetilde{\Phi_{\Lambda}})\,dV(x') \Rightarrow S'(L)$$

and (4.174) shows that $S(L)$ is isomorphic with $S'(L)$ and gives the transformation law induced by $\{\Phi_{\Lambda}\} \in \mathcal{G}(p,N;F_{\Lambda})/G_r$. However $S'(L)$ and $S(L')$ are trivially isomorphic since

$$\breve{J}[\{\Phi_{\Lambda}\}](L;D) \equiv \int_D L'\,dV(x) \equiv J[\{\Phi_{\Lambda}\}](L';D)\,.$$

The reader should note that (4.174) defines the *transformation properties* of $\{\mathcal{E}|L\}_{\Phi}$ that are induced by coordinate transformations of the arguments of $L$.

The results are now summarized. If we start with a given group, $G_r$, and a given $\mathfrak{D}_1(D;N)$, whose elements $\{\Phi_{\Lambda}(x^m)\}$ belong to some geometry class $\mathcal{G}(p,N;F_{\Lambda})/G_r$ (relative to $G_r$), then we can find the lagrangian class $L_c(G_r)$. If we take any $L \in L_c(G_r)$, then the functional $J[\{\Phi_{\Lambda}\}](L;D)$ is invariant under the action of $G_r$; that is, $J[\{\Phi_{\Lambda}\}](L;d) \equiv \breve{J}[\{\Phi_{\Lambda}\}](L;d) \equiv J[\{\Phi_{\Lambda}\}](L';d)$ for all $d \subset D$ and $L$ transforms as a scalar density under $G_r$. Of more importance, we have shown that if $\{\Phi_{\Lambda}\} \in S(L)$, then the image, $\{\Phi_{\Lambda'}\}$ of $\{\Phi_{\Lambda}\}$ under the transformation laws defined by $\{\Phi_{\Lambda}\} \in \mathcal{G}(p,N;F_{\Lambda})/G_r$ is an element of $S'(L) \equiv S(L')$: the transformed $\{\Phi_{\Lambda}\}$ is a solution of the transformed Euler equations. It is obvious from the theory of continuous $r$-parameter groups that the same results continue to hold for any subgroup of $G_r$ or for any group isomorphic to $G_r$. On the other hand, suppose that a group $G_k$ contains the group $G_r$ as a proper subgroup (obviously, $k > r$). The results then cease to hold, and we may not even know the transformation laws of $\{\Phi_{\Lambda}\}$ since it has been assumed only that $\{\Phi_{\Lambda}\} \in \mathcal{G}(p,N,F_{\Lambda})/G_r$ and not that $\{\Phi_{\Lambda}\} \in \mathcal{G}(p,N;F_{\Lambda})/G_k$. More explicitly, $(\partial_x\Phi_1)^2 - a^2(\partial_t\Phi_1)^2$ is invariant under the $G_3$ studied previously, while $g^{km}(\partial_k\Phi_1)(\partial_m\Phi_1)$, in which $g^{km}$ is a contravariant tensor under $G_k$, is invariant under the more general group $G_k$ containing $G_3$ as a proper subgroup.

*Problems*:

4.1 Show that $\mathcal{G}(p,N;F_\Lambda)/G_1$ is contained in $\mathcal{G}(p,N;F_\Lambda)/G_2$ only if $G_2$ is contained in $G_1$ (see Sec. 4.1).

4.2 Compute $\underset{v}{D}\Phi_\Lambda$, for $\{\Phi_\Lambda\}$ a scalar, a convariant vector, a contravariant vector, a covariant tensor of second order.

4.3 Compute $\left.\frac{dJ}{d\lambda}\right|_{\lambda=\eta=0}(d)$ and $\left.\frac{dJ}{d\pi}\right|_{\lambda=\eta=0}(d)$ for the general lagrangian (which gives rise to linear second-order systems) considered in Chap. 1.

4.4 (Research problem) Obtain the conditions for weak coordinate invariance when the lagrangian function is allowed to have functionals as arguments.

4.5 Obtain the equations which $v_a{}^m(x^k)$ must satisfy in order to have weak coordinate invariance for the functional whose Euler equations are (a) the three-dimensional wave equation, (b) equations for the complex scalar field, (c) equations for the charged, complex scalar field, (d) equations for the electromagnetic field.

4.6 The same as Prob. 4.5 with weak coordinate invariance replaced by weak point invariance.

4.7 Take $N = n = 2$, $\Phi = f_m$, where $\{f_m(x^k)\}$ is a covariant vector field on $X_2{}^u$, and $L = \frac{1}{2}(f_1)^2 + \frac{1}{2}(f_2)^2 - 3\lambda f_1 f_2 + \frac{1}{2}(\partial_1 f_2)^2 - (k^2/2)(\partial_2 f_1)^2 + \mu(\partial_1 f_1)(\partial_2 f_2)$.
Find the group for which we have weak coordinate invariance and the group for which we have weak point invariance in which $\partial_m \lambda = \partial_m \mu = \partial_m k = 0$.
Change $L$ by the addition of an element of $\mathfrak{N}(2)$ and find the groups which give weak coordinate invariance and weak point invariance. In particular, take an element of $A$ of $\mathfrak{N}(2)$ for which $\partial_m W_k{}^m(A) \neq 0$.

4.8 Work out the exact conditions for absolute coordinate invariance when (a) $N = 1$ and $\{\Phi_1\}$ is a scalar function, (b) $N = n$ and $\{\Phi_\Lambda\}$ is a covariant vector field, (c) $N = n$ and $\{\Phi_\Lambda\}$ is a contravariant vector field, (d) $N = n^2$ and $\{\Phi_\Lambda\}$ is a covariant tensor field of second order, (e) $N = n + 1$ and $\{\Phi_\Lambda\}$ consists of a covariant vector field and a scalar field. Take $n = 2$ and solve at least two of the above cases for $L$.

4.9 For each of the five cases given in Prob. 4.8, find the lagrangian functions which admit function invariance where either

$$H_\Lambda(\Phi_\Sigma, \mathcal{O}_\Lambda{}^1\psi_1) = e^{\psi_1}\Phi_\Sigma, \quad \mathcal{O}_\Lambda{}^1\psi_1 = e^{\psi_1}$$

or

$$H_\Sigma(\Phi_\Sigma, \mathcal{O}_\Lambda{}^1 \psi_1, \mathcal{O}_\Lambda{}^2 \psi_2) = e^{\psi_1} \Phi_\Sigma + K_\Sigma{}^m \partial_m \psi_2 ,$$

$$\mathcal{O}_\Lambda{}^1 \psi_1 = e^{\psi_1} , \quad \mathcal{O}_\Sigma{}^2 \psi_2 = K_\Sigma{}^m \partial_m \psi_2 ,$$

$$\partial_k K_\Sigma{}^m = 0 .$$

# Appendix

## STATIONARIZATION WITH CONSTRAINTS

A very important aspect of the calculus of variations has not been considered in the text, namely, the problem of stationarization when there are constraints present. The classical approach to such problems is somewhat combersome and fails to show a simple underlying method. This is due primarily to the fact that the classical calculus of variations does not consider functionals of functionals and hence is unable to consider integral constraints in a natural manner. If we follow the constructions given in Chap. 2, the ability to embed a given system of integrodifferential equations in a variational statement together with the stationarization of functionals of functionals provides a simple straightforward method of handling stationarization problems with constraints. In addition, the method to be used is the same in all cases—it does not depend on the form of the constraints—and gives the classicial results as special cases.

We use the notation introduced in Chap. 2 without additional explanation. Let

$$G_r = G_r[x^k, z^k, \Phi_\Lambda(z^k), \partial_m \Phi_\Lambda(z^k)]\ , \quad r = 1, \ldots, Q \tag{A.1}$$

be $Q$ given functions of class $C^2$ in their $2n + N + nN$ arguments, and form the $Q$ functionals

$$K_r(x^k; \Phi_\Lambda) = \int_{D^*} G_r\, dV(z)\ . \tag{A.2}$$

If

$$F_\beta = F_\beta[x^k, \Phi_\Lambda(x^k), \partial_m \Phi_\Lambda(x^k), K_r(x^k, \Phi_\Lambda)]\ , \quad \beta = 1, \ldots, P \tag{A.3}$$

are $P$ given functions of class $C^2$ in their $n+N+nN+Q$ arguments, we wish to stationarize the functional

$$J[\Phi_\Lambda](L) = \int_{D^*} L\,[x^k, \Phi_\Lambda(x^k), \partial_m \Phi_\Lambda(x^k), k_a(x^k; \Phi_\Lambda)]\, dV(x) \qquad (A.4)$$

with respect to those elements of $\mathfrak{D}_1(D^*; N)$ that satisfy the $P$ given constraints

$$F_\beta = 0\,. \qquad (A.5)$$

**Theorem A. 1.** *Define the quantity* $\tilde{L}$ *by the relation*

$$\tilde{L} = L + \lambda^\beta(x^k)F_\beta\,, \qquad (A.6)$$

*then* $J[\Phi_\Lambda](L)$ *is stationary with respect to those elements of* $\mathfrak{D}_1(D^*; N)$ *that satisfy the constraints* $F_\beta = 0$ *if and only if*

$$\tilde{J}[\Phi_\Lambda, \lambda^\beta](L) = \int_{D^*} \tilde{L}\, dV(x) \qquad (A.7)$$

*is stationary with respect to the elements of* $\mathfrak{D}_1(D^*; N+P)$ *constituted by the* $(N+P)$*-tuples of functions* $(\Phi_\Lambda(x^k), \lambda^\beta(x^k))$. *This is the case if and only if*

$$\{\mathcal{E}|\tilde{L}\}_{\lambda_\beta} = F_\beta = 0\,, \qquad (A.8)$$

$$\{\mathcal{E}|\tilde{L}\}_{\Phi_\Lambda} = \{\mathcal{E}|L\}_{\Phi_\Lambda} + \{\mathcal{E}|\lambda^\beta F_\beta\}_{\Phi_\Lambda} = 0 \qquad (A.9)$$

*hold throughout* $D^*$.

**Proof.** The constraints $F_\beta = 0$ are trivially embedded in a variational statement by the lagrangian function $\lambda^\beta(x^k)F_\beta$, since variation with respect to $\lambda$ yields the constraint equations. We are thus naturally led to consider the lagrangian function (A.6) on $\mathfrak{D}_1(D^*; N+P)$ in view of the linearity of the Euler-Lagrange operator with respect to lagrangian functions. If we stationarize $\tilde{J}$ with respect to $\lambda$ (i.e., we satisfy equations (A.8)), the constraint equations are satisfied. It thus follows from (A.6) and (A.8) that if $\tilde{J}$ is rendered stationary with respect to $\lambda$ then we have $\tilde{J} = J$ and hence we are then free to stationarize $J = \tilde{J}$ with respect to $\Phi$, which was our original stationarization problem. Stationarization with respect to $\Phi$ yields (A.9), and (A.8) and (A.9) are equivalent to stationarizing $\tilde{J}$ with respect to $\lambda$ and $\Phi$. Hence, $J$ is stationary with respect to all $\Phi$ that satisfy the constraints if and only if $\tilde{J}$ is stationary with respect to all elements of $\mathfrak{D}_1(D^*; N+P)$ and the result is established.

In order to see that Theorem A.1 contains the classical results as special cases, we must specialize the form of $F$ and $G$. Suppose that we have constraints defined by

$$F_\beta = \delta_\beta{}^r K_r(x^k;\Phi_\Lambda)\,, \tag{A.10}$$

$$G_r = G_r[z^k,\Phi_\Lambda(z^k),\partial_m\Phi_\Lambda(z^k)]\,, \tag{A.11}$$

so that

$$F_\beta = \delta_\beta{}^r K_r(\Phi_\Lambda) = \delta_\beta{}^r \int_{D^*} G_r[z^k,\Phi_\Lambda(z^k),\partial_m\Phi_\Lambda(z^k)]\,dV(z)\,. \tag{A.12}$$

We then have

$$\widetilde{L} = L + \lambda^\beta(x^k)\,\delta_\beta{}^r K_r(\Phi_\Lambda) = L + \lambda^r(x^k) K_r(\Phi_\Lambda)\,. \tag{A.13}$$

By Theorem A.1, we must then have

$$\{\mathcal{E}|\widetilde{L}\}_{\lambda^r} = K_r(\Phi_\Lambda) = \delta_\beta{}^r K_r(\Phi_\Lambda) = F_\beta = 0 \tag{A.14}$$

and

$$\{\mathcal{E}|\widetilde{L}\}_{\Phi_\Lambda} = \{\mathcal{E}|L\}_{\Phi_\Lambda} + \{\mathcal{E}|\lambda^r K_r\}_{\Phi_\Lambda} = 0\,. \tag{A.15}$$

Now, from (A.11), we have $G_r^* = G_r[x^k,\Phi_\Lambda(x^k),\partial_m\Phi_\Lambda(x^k)]$, while (A.13) yields $\partial(\lambda^r K_r)/\partial K_r = \lambda_r(x)$. Hence (see Chap. 2) we have

$$\{\mathcal{E}|\lambda^r K_r\}_{\Phi_\Lambda} = \{e|\lambda^r K_r\}_{\Phi_\Lambda} + \int_{D^*}\frac{\partial(\lambda^r K_r)}{\partial K_r}(z)\,\{e|G_r^*\}_{\Phi_\Lambda}\,dV(z)$$

$$= \int_{D^*}\lambda_r(z)\,\{e|G_r^*\}_{\Phi_\Lambda}\,dV(z) = \{e|G_r^*\}_{\Phi_\Lambda}\int_{D^*}\lambda^r(z)\,dV(z)\,,$$

since $\{e|G_r^*\}$ is independent of $z^k$ in this case. Setting $\overline{\lambda}^r = \int_{D^*}\lambda^r(z)\,dV(z)$, which are constants, we have that

$$0 = \{\mathcal{E}|\widetilde{L}\}_{\Phi_\Lambda} = \{\mathcal{E}|L\}_{\Phi_\Lambda} + \overline{\lambda}^r\{e|G_r^*\}_{\Phi_\Lambda} = \{\mathcal{E}|L + \overline{\lambda}^r G_r^*\}_{\Phi_\Lambda}\,.$$

This is, however, precisely the classical result when there are integral constraints, (A.12), to be satisfied. The student should note that we did not have to assume that $\overline{\lambda}$ was a constant. This results automatically from the method since $\bar{\lambda}$ turns out to be the definite integral of $\lambda(x)$ over the given domain $D^*$. If $F$ is independent of $K$, then Theorem A.1 gives exactly the classical results in that case; and $\lambda$ is then a function of $x$.

Theorem A.1 is obviously more widely applicable than the classical theorems and includes the cases in which $\lambda$ satisfies systems of integrodifferential equations if there is to be a stationarization. As an example, suppose that

$$F_\beta = \mathcal{O}_{\beta\Lambda}\Phi_\Lambda - f_\beta(x^k), \tag{A.16}$$

where $\mathcal{O}_{\beta\Lambda}$ is a linear operator of second order. The results of Chap. 2 then give us, in conjunction with Theorem A.1,

$$\widetilde{L} = L + \lambda^\beta(x)[\mathcal{O}_{\beta\Lambda}\Phi_\Lambda - f_\beta(x^k)], \tag{A.17}$$

$$\{\mathcal{E}|\widetilde{L}\}_{\lambda\beta} = \mathcal{O}_{\beta\Lambda}\Phi_\Lambda - f_\beta(x^k) = 0, \tag{A.18}$$

$$\{\mathcal{E}|\widetilde{L}\}_{\Phi_\Lambda} = \{\mathcal{E}|L\}_{\Phi_\Lambda} + \{\mathcal{E}|\lambda^\beta\mathcal{O}_{\beta\Lambda}\Phi_\Lambda\}_{\Phi_\Lambda} = \{\mathcal{E}|L\}_{\Phi_\Lambda} + \mathcal{O}^+_{\beta\Lambda}\lambda^\beta = 0, \tag{A.19}$$

where $\mathcal{O}^+_{\beta\Lambda}$ is the adjoint of $\mathcal{O}_{\beta\Lambda}$. We thus have the system of equations

$$\mathcal{O}_{\beta\Lambda}\Phi_\Lambda - f_\beta = 0, \quad \mathcal{O}^+_{\beta\Lambda}\lambda^\beta + \{\mathcal{E}|L\}_{\Phi_\Lambda} = 0 \tag{A.20}$$

to solve for $[\Phi_\Lambda(x^k), \lambda^\beta(x^k)]$ if $J[\Phi_\Lambda](L)$ is to be stationary for $\Phi_\Lambda(x)$ satisfying $\mathcal{O}_{\beta\Lambda}\Phi_\Lambda = f_\beta$

As a further example, consider the case where $P = Q = 1$,

$$F_1 = K_1, \quad G_1 = G_1[x^k, z^k, \Phi_\Lambda(z^k), \partial_m\Phi_\Lambda(z^k)]. \tag{A.21}$$

We then have

$$\begin{aligned}\widetilde{L} &= L + \lambda(x^k)\int_{D^*} G_1[x^k, z^k, \Phi_\Lambda(z^k), \partial_m\Phi_\Lambda(z^k)]\,dV(z)\\ &= L + \lambda K_1,\end{aligned} \tag{A.22}$$

$$\{\mathcal{E}|\widetilde{L}\}_\lambda = F_1 = K_1 = \int_{D^*} G_1\,dV(z) = 0, \tag{A.23}$$

$$\begin{aligned}\{\mathcal{E}|\widetilde{L}\}_{\Phi_\Lambda} &= \{\mathcal{E}|L\}_{\Phi_\Lambda} + \{\mathcal{E}|\lambda K_1\}_{\Phi_\Lambda}\\ &= \{\mathcal{E}|L\}_{\Phi_\Lambda} + \int_{D^*}\lambda(z^k)\{e|G_1^*\}_{\Phi_\Lambda}\,dV(z) = 0.\end{aligned} \tag{A.24}$$

In this case, since

$$G_1^* = G_1[z^k, x^k, \Phi_\Lambda(x^k), \partial_m \Phi_\Lambda(x^k)]$$

depends on $z^k$, $\{e|G_1^*\}$ cannot be taken outside of the $z$-wise integration and hence $\lambda(x)$ satisfies the $N$ integral equations (A.24). The reader should note that equations (A.9) *are always linear equations in* $\lambda$ and hence linear theory can be used to obtain $\lambda$.

# List of Frequently Used Symbols

| | |
|---|---|
| $A_n$ | — $n$-dimensional number space |
| $\mathcal{C}(D^*; N)$ | — space of all $N$-tuples of continuous functions on $D^*$ with the uniform convergence norm $\mathcal{C}(D^*) = \mathcal{C}(D^*; 1)$ |
| $D$ | — open, connected point set in $n$-dimensional space with compact closure and nonzero volume measure |
| $D^*$ | — closure of $D$ |
| $\partial D$ | — boundary of $D^*$ |
| $\underset{v}{D}$ | — $\underset{v}{D}\phi = v^k \partial_k \phi - \underset{v}{\pounds}\phi$ |
| $\mathfrak{D}_1(D^*; N)$ | — space of $N$-tuples of functions with continuous first partial derivatives in $D^*$ and with the uniform convergence norm on the functions and their derivatives $\mathfrak{D}_1(D^*) = \mathfrak{D}_1(D^*; 1)$ |
| $d$ | — any connected subset of $D$ with nonzero volume measure |
| $\{\mathcal{E}\|L\}$ | — Euler-Lagrange operator (determined by $L$) |
| $\{e\|L\}$ | — local Euler-Lagrange operator (determined by $L$) |
| $F_\Lambda$ | — transformation functions for geometric objects |
| $G_r$ | — $r$-parameter continuous group |
| $G_n{}^u$ | — pseudogroup of coordinate transformations of class $C^u$ on $X_n{}^u$ |
| $\mathcal{G}(p; N; F_\Lambda)/G$ | — geometry class relative to the group $G$ |
| $k_a(x^k;\ )$ | — an indexed set of functionals with parameters $x^k$ |
| $L$ | — lagrangian function |
| $\hat{L}$ | — equivalence class of lagrangian functions with the same Euler equations |
| $L_c(G_r)$ | — lagrangian coordinate class determined by $G_r$ |
| $L_p(G_r)$ | — lagrangian point class determined by $G_r$ |
| $\mathfrak{N}(N)$ | — null class of the Euler-Lagrange operator when the function space consists of $N$-tuples of functions $\mathfrak{N} = \mathfrak{N}(1)$ |

| | |
|---|---|
| $\mathfrak{N}^*(N)$ | — the trivial null class of the Euler-Lagrange operator when the function space consists of $N$-tuples of functions $\mathfrak{N}^* = \mathfrak{N}^*(1)$ |
| $\mathcal{O}$ | — operator (usually integrodifferential and linear) |
| $\mathcal{O}^+$ | — adjoint of $\mathcal{O}$ |
| $S(L)$ | — set of all elements of the function space which stationarize a functional that is the integral of $L$ over $D^*$ |
| $dS_i(x)$ | — differential element of directed surface area of $\partial D^*$ with respect to the $x$-coordinate system |
| $dV(x)$ | — differential element of volume of $D^*$ with respect to the $x$-coordinate system |
| $W_k{}^m(L)$ | — momentum-energy complex formed from $L$ |
| $X_n$ | — $n$-dimensional space |
| $X_n{}^u$ | — $n$-dimensional manifold of class $u$ |
| $x^k$ | — coordinates of a point with respect to the $x$-coordinate system |
| $\underset{v}{\pounds}$ | — Lie derivative with respect to $v^k(x^m)$ |
| $\partial_m$ | — total partial derivative with respect to the $m$th independent variable |
| $\partial^*_m$ | — explicit partial derivative $\partial_m F[x^k, y(x^k)] = \partial^*_m F + (\partial F/\partial y)\partial_m y$ |
| $\underset{v}{\Delta}$ | — total variation |

# Subject Index